# ATLAS

## OF SCIENCE LITERACY

VOLUME 2

# ATLAS

## of Science Literacy

Volume 2

**PROJECT 2061**

American Association for the Advancement of Science
National Science Teachers Association, co-publisher

Washington, DC
2007

Library of Congress Cataloging-in-Publication Data
Atlas of science literacy / American Association for the Advancement of Science, Project 2061.
p. cm.
Includes bibliographical references and index.
ISBN 978-0-87168-712-8
1. Science—Study and teaching—United States—Charts, diagrams, etc. 2. Mathematics—Study and teaching—United States—Charts, diagrams, etc. 3. Engineering—Study and teaching—United States—Charts, diagrams, etc. I. Project 2061 (American Association for the Advancement of Science)

ISBN-13: 978-0-87168-712-8
ISBN-10: 0-87168-712-7

Book design by Kessler Design Group

1 3 5 7 9 8 6 4 2
Printed in the United States of America

This material is based upon work supported by the National Science Foundation under Grant No. ESI-0103678.

Any opinions, findings, and conclusions or recommendations expressed in this material are those of the author(s) and do not necessarily reflect the views of the National Science Foundation.

B

# TABLE OF CONTENTS

The **American Association for the Advancement of Science** (AAAS) is the world's largest general scientific society and publisher of the journal *Science* (www.sciencemag.org). AAAS was founded in 1848, and serves 262 affiliated societies and academies of science, reaching 10 million individuals. *Science* has the largest paid circulation of any peer-reviewed general science journal in the world, with an estimated total readership of 1 million. The non-profit AAAS (www.aaas.org) is open to all and fulfills its mission to advance science and serve society through initiatives in science policy, international programs, science education, and more. For the latest research news, log onto EurekAlert! (www.eurekalert.org), the premier science-news Web site, a service of AAAS.

A long-term initiative of AAAS, **Project 2061** began its work in 1985—the year Halley's Comet was last visible from Earth. Children starting school then and now will see the return of the Comet in 2061—a reminder that today's education will shape the quality of their lives as they come of age in the 21st century amid profound scientific and technological change.

Project 2061's mission is to help all Americans become literate in science, mathematics, and technology. Its 1989 report *Science for All Americans* defines what that literacy means and identifies the reforms needed to achieve it. Today, Project 2061 (www.project2061.org) conducts research and develops tools and services that educators, researchers, and policymakers can use to make critical and lasting improvements in the nation's educational system. The Project's areas of expertise include:

**Curriculum Evaluation & Design.** High-quality content standards are essential to efforts to improve student achievement in science and mathematics. Through its *Benchmarks for Science Literacy* and *Atlas of Science Literacy, Volume 1* and *Volume 2*, Project 2061 has provided a set of learning goals that states can use as a foundation for their own standards. To help educators identify classroom resources that are aligned with these goals, Project 2061 has evaluated the most widely used middle- and high-school science and mathematics textbooks and found only a few that were likely to be useful. With funding from the National Science Foundation (NSF), Project 2061 is now leading the Center for Curriculum Materials in Science to conduct research and develop new leadership that will result in a new generation of more effective materials for science teaching and learning. Project 2061 staff also develops model components for curriculum design and consults extensively with curriculum developers and publishers.

**Assessment.** As statewide science testing becomes part of the No Child Left Behind mandate, the alignment of assessment items with national and state science and mathematics content standards is more important than ever. To help test developers, education researchers, and teachers deal with this challenge, Project 2061 is creating an online bank of high-quality test items for use in middle- and early high-school science and mathematics. Each item in the NSF-funded bank has been screened for alignment using criteria developed through Project 2061 research and is suitable for use with a diverse range of students, including English-language learners. Project 2061 has also been providing its expertise in the development of the science framework for the 2009 National Assessment of Educational Progress (NAEP).

**Teacher Development.** Improving student achievement in science and mathematics requires well-prepared teachers who have a solid grounding in science and mathematics and an understanding of how diverse students learn. Project 2061 is involved in a range of research and development efforts aimed at improving teacher knowledge and skills and providing them with instructional resources focused on benchmarks and standards. Findings from a Project 2061 study of middle-school mathematics teachers and their students are likely to shed light on how textbooks, teaching practices, and professional development can best be coordinated to improve student learning. Project 2061's conceptual strand maps in *Atlas of Science Literacy, Volume 1* and *Volume 2* are providing the interface for several digital library projects and are the focus of several professional development workshops each year.

AAAS wishes to express its gratitude to the **National Science Foundation** for its generous support of Project 2061.

The **National Science Teachers Association** (NSTA), founded in 1944 and headquartered in Arlington, Virginia, is the largest organization in the world committed to promoting excellence and innovation in science teaching and learning for all. NSTA's current membership of more than 55,000 includes science teachers, science supervisors, administrators, scientists, business and industry representatives, and others involved in and committed to science education. For more information, visit www.nsta.org.

Project 2061 and the National Science Teachers Association jointly publish *Atlas of Science Literacy, Volume 1* and *Volume 2*.

# FOREWORD

In 1990 I had the honor of joining the National Council on Science and Technology Education, which had been established by AAAS to help guide its ambitious new science education initiative Project 2061. The Council's discussions often focused on the need for materials and ideas that would help educators realize the vision of science literacy characterized in Project 2061's groundbreaking 1989 report *Science for All Americans*.

Within a few years, *Benchmarks for Science Literacy* was developed to help define the concepts and skills in science, mathematics, and technology that students would need to learn at each grade level as they made progress toward science literacy. Although *Benchmarks* emphasized the interconnectedness of that knowledge and the importance of building on what students learn as they move from kindergarten through high school, it was difficult to make those relationships among ideas and the progression of learning from grade to grade readily apparent to readers. What was needed was a visual representation of the connections among ideas from one grade band to the next and within a grade band. The solution was a series of conceptual strand maps.

In 1997 the National Science Foundation funded the development of the first set of maps. This set, which became known as *Atlas of Science Literacy*, was co-published with NSTA in 2001. We fondly called *Atlas* a coffee table book of science literacy because of its size, and we pictured it in every living room in the nation. Although the first volume of *Atlas* did not cover all of the K-12 learning goals in *Benchmarks*, it was a productive start. School systems used the book to reconsider their entire curricula; individual teachers and science department faculty found that *Atlas* could help in furthering their own science learning and in promoting more coherence in their science curriculum; assessment developers used maps to help align test items with grade levels and topics; preservice methods instructors used maps to help those entering the profession to analyze curriculum content, design, and effectiveness; and the list goes on. The success of the first volume of *Atlas* has resulted in over 32,000 copies in print, with orders from around the globe, and the development of highly successful workshops conducted throughout the United States. The *Atlas* is widely cited in the science education literature as an essential tool for reform. *Atlas* maps provide the much-needed "big picture" perspective; their value is undeniable.

*Atlas of Science Literacy, Volume 2* completes the collection of maps. Together *Atlas 1* and *Atlas 2* cover all of the K-12 learning goals recommended in *Benchmarks for Science Literacy*. The release of this second volume is particularly timely and coincides with a renewed interest in science education spurred, at least in part, by statewide science testing mandated under federal No Child Left Behind legislation. *Atlas 2* comes at a time when we, as teachers, find support for what we are doing, both as professionals and as concerned citizens. Both *Atlas* volumes serve as remarkable resources for current efforts to improve curriculum, instruction, and assessments. NSTA is proud to be involved in the production of this important addition to your library; *Atlas 2* promotes excellence and innovation in science teaching and learning for all.

Linda Froschauer
NSTA President 2006-2007
8th Grade Science Teacher
Weston Public Schools, Connecticut

*This publication is dedicated to*
*Andrew (Chick) Ahlgren (1936-2006),*
*who brought map making and cartoons—*
*like his own cartoon above—to Project 2061*
*and changed forever how we think about*
*science learning and teaching.*

# PREFACE

A central premise of today's standards-based approach to science education is that science literate people possess knowledge that is richly interconnected, forming a coherent understanding of the world and how it works. This premise has shaped Project 2061's efforts to change science education profoundly, so that rather than knowing fragmented bits and pieces of information, students understand the most important ideas, how those ideas fit together, and how to apply them in a variety of contexts. Thinking about these connections and how to describe and portray them effectively has been integral to Project 2061's work since the beginning.

Project 2061's first publication *Science for All Americans* presents a coherent story of what science literate adults should know and be able to do. Deciding on these literacy goals often involved imagining what the conceptual steps would be and whether they were feasible for all students. When we then set out to identify in *Benchmarks for Science Literacy* what students need to learn from kindergarten through 12th grade to achieve those adult literacy goals, attention to connections among benchmarks was integral to that effort. With the first volume of *Atlas of Science Literacy*, we have found that the visual appeal of seeing how ideas relate to each other on the maps helps educators to appreciate the importance of their conceptual and developmental connections. In this new *Atlas* volume, the maps are even more explicit about relationships among ideas within and across topic areas and grade ranges.

In preparing the new *Atlas* maps for publication, we have come to realize how significant the maps—and the process of mapping itself—have become to our work and to the work of many others. Here at Project 2061, we've created specialized maps to analyze the coherence of textbooks' treatment of key science ideas and to support our analysis and development of assessment items aligned to benchmarks and standards. Curriculum materials developers with whom we work use maps to guide the design and sequencing of activities. And our digital library collaborations have shown the promise of interactive strand maps for accessing standards-based resources online.

The 44 strand maps in this second volume complete the set of some 100 maps derived from early maps that were in the minds of the authors of *Benchmarks for Science Literacy*. Readers will note the many revisions in the text of benchmarks on the maps and, to a lesser extent, in the grade-level placement of benchmarks. These changes update the science, take account of recent learning research, and reflect our own experience and that of educators in using benchmarks to inform the analysis and design of curriculum and assessment. Similar changes to maps in *Atlas 1* are planned. Our hope is that the maps in *Atlas 1* and *Atlas 2* will continue to enrich the efforts of everyone dedicated to science literacy for all, and we look forward to hearing about this work.

Jo Ellen Roseman
Director, Project 2061
American Association for the Advancement of Science

# ATLAS

## OF SCIENCE LITERACY

VOLUME 2

Comet Halley, March 1986. Anglo-Australian Observatory/Royal Observatory, Edinburgh

# INTRODUCTION

Science, mathematics, and technology have a profound impact on our individual lives and our culture. They play a role in almost all human endeavors, and they affect how we relate to one another and the world around us. Project 2061 is a long-term effort to improve education so that all citizens attain science literacy—that is, an understanding of the natural and social sciences, mathematics, technology, and their interactions. Science literacy enables us to make sense of real-world phenomena, informs our personal and social decisions, and serves as a foundation for a lifetime of learning.

Project 2061's seminal report *Science for All Americans* (American Association for the Advancement of Science [AAAS], 1989) provides a narrative account of the knowledge and abilities that make up adult science literacy. Its companion publication *Benchmarks for Science Literacy* (AAAS, 1993) lists specific ideas and skills that serve as goals for student learning in four grade ranges—steps along the way to achieving literacy. *Science for All Americans* and *Benchmarks*, together with other standards documents, are being used nationwide to reshape state and local curriculum frameworks, instruction, assessment, and teacher education. While the learning goals proposed in these documents are necessary for making changes in science education, they are not sufficient for ensuring that those changes are coordinated to move all students toward science literacy.

*Atlas of Science Literacy* attends to the critical challenge of ensuring the coherence of science education reforms: Science literacy should be approached not as a compilation of isolated bits of information and abilities, but as a rich fabric of mutually supporting ideas and skills that must develop over time. From kindergarten through high school, what students learn should build on what they have learned before, make sense in terms of what else they are learning, and prepare them for what they will learn next. To help students achieve science literacy, educators need to see how the ideas and skills that students learn in different grades and topics—and even disciplines—depend on and support one another. *Atlas* depicts this pattern of connections in a set of "strand maps" that provide a graphic representation of students' growth of understanding. Each map displays the ideas, skills, and connections among them that are part of achieving literacy in a particular topic, showing where each step along the way comes from and where it leads.

*Atlas* is intended to help educators understand what students can be expected to learn in different grades and to help foster the design of more coherent curriculum, instruction, and assessment. Unless these three elements of science education promote the development of a coherent set of ideas and skills over time, students may be left with nothing more than unrelated, poorly understood, and quickly forgotten facts, algorithms, and technical terms.

The maps in *Atlas* do not prescribe a particular curriculum or instructional strategy. Instead, they present a framework meant to inspire a variety of different ways to design and organize learning experiences suited to local circumstances. Furthermore, these maps will always be open to improvement in light of new discussions, classroom experiences, and research on learning. We hope the enduring message of *Atlas* will be that thinking carefully about the growth of understanding from kindergarten through high-school graduation is an essential part of planning what students can be expected to learn and how best they can be helped to do so.

## ABOUT *ATLAS OF SCIENCE LITERACY*

*Atlas of Science Literacy* is a two-volume collection of nearly 100 conceptual strand maps—and commentary on those maps—that show how students' understanding of the ideas and skills that lead to literacy in science, mathematics, and technology might develop from kindergarten through 12$^{th}$ grade.

*Atlas* is part of a coordinated set of tools developed by Project 2061 to help educators understand and use specific goals for student learning. The maps in *Atlas* are built from the K-12 learning goals—or benchmarks—presented in *Benchmarks for Science Literacy*. *Benchmarks* itself was derived from the recommendations for adult science literacy proposed in *Science for All Americans*. In addition, the connections among benchmarks that are depicted in the maps are based on the cognitive research and essays in *Benchmarks*, and on the account of adult literacy in *Science for All Americans*.

Volume 1 of *Atlas* was published in 2001 and is joined now by a second volume; together they map all of the goals that are recommended in *Benchmarks* as essential for every student to learn. Unless otherwise noted, the following description of the content of *Atlas* and suggestions for using the maps and other features pertain to both Volume 1 and Volume 2.

## USING *ATLAS*

The maps in *Atlas* enhance the content of *Benchmarks* (and other standards documents) by displaying explicit connections among individual benchmarks.

Because these connections are themselves a critical part of science literacy—and of helping students achieve it—we have found the maps invaluable for guiding our own research and development. Since the publication of Volume 1, educators working in a wide range of settings have made extensive use of the maps to carry out their various responsibilities. The examples below illustrate just a few of the ways maps are being used to improve science education.

*Understanding benchmarks and standards.* By studying maps carefully, teachers and other educators enhance their own understanding of the content and nature of the benchmarks as specific learning goals. In Project 2061's professional development workshops, for example, maps help educators clarify the meaning of individual benchmarks and get a sense of what the benchmarks in a strand are trying to achieve. Of course, local frameworks or other sets of standards can also be studied and compared to the maps to shed light on their coherence and comprehensiveness—or lack thereof.

*Designing curriculum.* The information in the maps helps educators distribute responsibilities for students' science learning across different grades and subjects, thus fostering the cumulative effectiveness of instruction. The assignment of benchmarks to different grades and courses should take account of connections among the ideas and skills students will be expected to learn. Several states and districts are using maps to inform this process.

*Planning instruction.* Maps enable educators to prepare instruction that is focused on the specific ideas in a benchmark and to take account of the precursors these specific ideas build on. When students are struggling with an idea, maps are used to decide what conceptual support might be needed or helpful. Learning experiences that take advantage of the connections among benchmarks are more likely to foster understanding. In the long run, helping students see how the things they learn fit together enhances their understanding and retention.

*Developing or evaluating curriculum materials.* Maps offer materials developers a helpful perspective on which benchmarks to target and when to target them. Maps also help them think about activities and lessons that address the specific substance of the benchmarks—rather than only something under the same general topic heading—at an appropriate level of sophistication. A critical step in either developing or analyzing materials is to gain a deep understanding of the targeted benchmark itself, the prerequisite ideas that contribute to it, and conceptual connections that should be highlighted in the student text or the teacher guide. In Project 2061's evaluation of biology textbooks, for example, analysts used maps to examine and keep track of whether textbooks presented a coherent set of ideas or presented ideas piecemeal, leaving out critical parts of the story (see www.project2061.org/publications/textbook/hsbio/report/browse.htm). Several materials development projects, particularly at the middle-school level, are using maps to guide their thinking and inform their designs.

*Constructing and analyzing assessment.* When used in designing and interpreting assessment, maps help answer questions about when it is appropriate to assess particular ideas and skills, and why students might have had trouble with a particular task. With maps as a framework, assessments are being designed to probe where students are in their growth of understanding, rather than how well they know a collection of unrelated ideas. In Project 2061's analysis of assessments, maps continue to be a key tool in deciding which ideas in a benchmark are actually needed to complete a task.

*Preparing teachers.* Whether in a pre-service or in-service context, all of the activities listed above can sharpen teachers' sense of what benchmarks mean and how to help students attain them. Teachers report that their own content knowledge is improved by studying the connections among benchmarks both within and across grade ranges. Maps portray literacy in science, mathematics, and technology from the viewpoint of how students are likely to develop that knowledge, providing a perspective on the content that is helpful to educators in thinking about how to teach it.

*Organizing resources. Atlas* strand maps are proving to be useful frameworks for organizing education resources and linking them to particular ideas that are found in national and state science standards. For example, several large digital library collections use electronic versions of maps as their online interface; teachers and other users are able to "browse" a map and access the resources that are linked to it by clicking on specific benchmarks.

## THE SET OF MAPS

The publication of Volume 2 completes the set of maps begun in Volume 1 of *Atlas*. Together, the two volumes map all of the learning goals specified in *Benchmarks*. In addition to contributing new maps to *Atlas* chapters that were not fully covered in Volume 1, Volume 2 also introduces maps for two chapters that were not covered at all in the earlier volume. Chapter 10: HISTORICAL PERSPECTIVES introduces nine new maps on important episodes of scientific discovery, and Chapter 12: HABITS OF MIND presents six maps that deal with attitudes, skills, and values that are essential to science literacy.

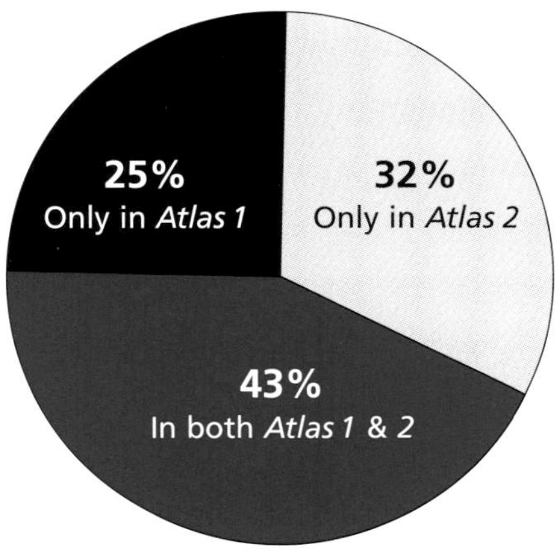


A benchmark may appear only on maps in *Atlas 1*, only on maps in *Atlas 2*, or on maps in both volumes.

To make it easier for readers to access the complete set of maps, the Index of Mapped Benchmarks beginning on page 124 covers all of the maps published in both volumes of *Atlas*. In addition, the inside front cover of this volume presents a combined list of all maps and their page numbers organized by *Benchmarks* chapter and section, and the inside back cover presents a list of all maps in both volumes in alphabetical order. To emphasize the full coverage of *Benchmarks*, each chapter begins with a combined list of the maps for that chapter published in both volumes. Maps in Volume 2 also identify references to maps published in Volume 1 with "v. 1" followed by the page number (see the Map Key on page *xxii*).

## ORGANIZATION OF *ATLAS*

The chapter organization of *Atlas* follows that used in both *Science for All Americans* and *Benchmarks*. Each *Atlas* chapter includes maps that correspond to the sections in the matching *Benchmarks* chapters. But because the *Atlas* maps juxtapose benchmarks from several different *Benchmarks* chapters and sections, maps named for a particular section in *Benchmarks* may include contributing benchmarks from other chapters or sections.

Indeed, one value of the maps is that they bring together benchmarks relevant to each topic regardless of where they originally appeared in *Benchmarks*. For example, the **USE OF EARTH'S RESOURCES** map in Chapter 4: THE PHYSICAL SETTING draws benchmarks from two sections of that chapter but also includes benchmarks from Chapter 5: THE LIVING ENVIRONMENT and Chapter 8: THE DESIGNED WORLD.

Although we have tried to align the set of maps in *Atlas* and the organization of *Benchmarks* as closely as possible, the process of putting benchmarks into maps inevitably gives rise to a new organization and challenges the reader to think about K-12 science learning goals in a new way.

## MAP COMMENTARY

Each *Atlas* map is accompanied by commentary on the facing page. The first part of the commentary provides a general discussion of the topic at hand, a brief summary of the content of the map and its major strands, and remarks on the focus of learning at each of the four grade ranges. In some cases, particularly interesting or important links to related maps are also pointed out. Maps published in *Atlas 1* are identified as such; all other maps are in this volume.

The second part of the map commentary includes "Notes" and "Research" relevant to the maps. The notes are not intended to provide an exhaustive account of the ideas in the map but rather to point out aspects of the map that may be of interest but are not readily apparent to the reader. For example, the notes might comment on an idea not currently found in *Benchmarks* that has been added to the map to strengthen or make explicit a logical or cognitive connection or on instances when a benchmark has been moved from one grade range to another. Also noted are benchmarks that play a particularly important purpose in one or more strands or that play a key role in other maps. Relevant cognitive research is also included where it is available. In most cases, this research has been excerpted from the summaries in *Benchmarks* Chapter 15: THE RESEARCH BASE and is limited to the research that was available at the time *Benchmarks* was published. Only those references that are particularly helpful in clarifying the map are presented. For a few maps in Volume 2, research findings from other sources have been included to update or supplement the research reported in *Benchmarks*.

## ABOUT STRAND MAPS

An *Atlas* strand map focuses on a topic important for literacy in science, mathematics, and technology and displays the benchmarks—from kindergarten through high school—that are most relevant to understanding it, suggesting for each benchmark along the way earlier benchmarks it builds on and later benchmarks it supports. A central feature of the maps is the development over time of the ideas and skills expressed in the benchmarks. To highlight the developmental nature of the maps, we call them *strand maps*. By *strand* we mean identifiable concepts or stories that are developed in sets of benchmarks across different grade levels. But sorting the benchmarks in a map into distinct, rigidly defined strands is not necessarily helpful or even tenable. The concepts and stories that make up literacy often interweave, join together, and share benchmarks as they develop.

### BENCHMARKS

The ideas and skills presented in the maps are specific goals for student learning. We call these learning goals *benchmarks* because they are derived from *Benchmarks for Science Literacy*, in which each benchmark is prefaced by the stem "Students should know that..." or the stem "Students should be able to...." Of course, mapping could be attempted with any reasonably specific, coherent, and complete set of learning goals. Many of the issues raised and lessons learned in mapping benchmarks should also apply to other sets of learning goals.

Attention to connections among ideas and skills was a critical part of developing *Benchmarks*, but because each benchmark might be involved in multiple contexts, it often could not be written to relate optimally to every other relevant benchmark. Making explicit connections among benchmarks in the maps inevitably has drawn more acute attention to their relationships and has suggested some fine-tuning. For example, many benchmarks contain several related ideas or skills. In the maps, these are often separated to clarify which specific ideas within one benchmark connect to other specific ideas in another benchmark. Figure 1 shows how a single goal in *Benchmarks* appears as two separate boxes on the maps.

The development of an *Atlas* map requires careful consideration of each benchmark that is to be included. As a result, some benchmarks have been reworded to clarify the flow of ideas and skills in the map. In other instances, the close scrutiny involved in mapping has prompted more significant changes. Despite these modifications, the benchmarks in the maps are consistent with the original versions as they appear in *Benchmarks*.

In some cases, ideas or skills have been placed at grade levels higher or lower than in *Benchmarks*. In other cases, an additional contributing idea or skill was needed. Such needs have been met mostly by including an idea from *Science for All Americans* (see Figure 2 for an example) or the essays in *Benchmarks*, but, in a few cases, a new benchmark has been constructed or adapted from *National Science Education Standards* (National Research Council, 1996). Overall, the process of mapping has proved to be an effective way to fine-tune *Benchmarks*.

To help readers correlate the text on the maps with the goals in *Benchmarks* and to point out where changes and additions have been made, the text of each mapped benchmark is followed by a code that references the chapter, section, grade range, and number of the corresponding bulleted goal statement in *Benchmarks*. In cases where only part of a benchmark is included on the map, lowercase letters in the code indicate where in the benchmark it comes from (e.g., "a" indicates the first sentence in the benchmark, "b" indicates the second sentence, and so on). A single asterisk at the end of the code indicates that the benchmark has been edited; two asterisks signal that the idea on the map is not currently included as a bulleted goal in *Benchmarks* but is derived instead from statements in *Science for All Americans* and *National Science Education Standards*, from the essays or other elements of *Benchmarks*, or (rarely) from Volume 1 of the *Atlas*. (See Map Key on page *xxii* for a full explanation of the coding scheme.) Readers can find all of the changes that have been made to individual benchmarks

Figure 1: Text from *Benchmarks* and boxed benchmarks on the maps.

*Benchmarks for Science Literacy*
CHAPTER **1** THE NATURE OF SCIENCE
Section 1A The Scientific World View

**Grades 6 through 8**

Most early adolescents have a more immediate interest in nature than in the philosophy of science. They should continue to be engaged in doing science and encouraged to reflect on the science they are engaged in, with the assumption that they will later acquire a more mature reflection on science as a world view.

Early adolescence, however, is not too early to begin to deal with the question of the durability of scientific knowledge, and particularly its susceptibility to change. Both incremental changes and more radical changes in scientific knowledge should be taken up. Radical changes in science sometimes result from the appearance of new information, and sometimes from the invention of better theories (for example, germ theory and geologic time, as discussed in Chapter 10: HISTORICAL PERSPECTIVES).

By the end of the 8th grade, students should know that

- When similar investigations give different results, the scientific challenge is to judge whether the differences are trivial or significant, and it often takes further studies to decide. Even with similar results, scientists may wait until an investigation has been repeated many times before accepting the results as correct.
- Scientific knowledge is subject to modification as new information challenges prevailing theories and as a new theory leads to looking at old observations in a new way.
- Some scientific knowledge is very old and yet is still applicable today.
- Some matters cannot be examined usefully in a scientific way. Among them are matters that by their nature cannot be tested objectively and those that are essentially matters of morality. Science can sometimes be used to inform ethical decisions by identifying the likely consequences of particular actions but cannot be used to establish that some action is either moral or immoral.

When similar investigations give different results, the scientific challenge is to judge whether the differences are trivial or significant, and it often takes further studies to decide. 1A/M1a

Even with similar results, scientists may wait until an investigation has been repeated many times before accepting the results as correct. 1A/M1b

Figure 2: Text from *Science for All Americans* and boxed benchmarks on the maps.

*Science for All Americans*
CHAPTER **1** THE NATURE OF SCIENCE
Section 1B Scientific Inquiry

**Science Is a Blend of Logic and Imagination**
Although all sorts of imagination and thought may be used in coming up with hypotheses and theories, sooner or later scientific arguments must conform to the principles of logical reasoning—that is, to testing the validity of arguments by applying certain criteria of inference, demonstration, and common sense. Scientists may often disagree about the value of a particular piece of evidence, or about the appropriateness of particular assumptions that are made—and therefore disagree about what conclusions are justified. But they tend to agree about the principles of logical reasoning that connect evidence and assumptions with conclusions.

Scientists do not work only with data and well-developed theories. Often, they have only tentative hypotheses about the way things may be. Such hypotheses are widely used in science for choosing what data to pay attention to and what additional data to seek, and for guiding the interpretation of data. In fact, the process of formulating and testing hypotheses is one of the core activities of scientists. To be useful, a hypothesis should suggest what evidence would support it and what evidence would refute it. A hypothesis that cannot in principle be put to the test of evidence may be interesting, but it is not likely to be scientifically useful.

To be useful, a hypothesis should suggest what evidence would support it and what evidence would refute it. A hypothesis that cannot, in principle, be put to the test of evidence may be interesting, but it may not be scientifically useful. 1B/H9** (SFAA)

in this volume reflected in the online version of *Benchmarks* (www.project2061.org/publications/bsl/online/bolintro.htm). The online version also incorporates the benchmark codes used for the Volume 2 maps.

## CONNECTIONS

Connections between benchmarks are based on the logic of the subject matter and, insofar as possible, on the published research into how students learn—both in general and with regard to specific concepts. A connection between two benchmarks, represented in the maps by an arrow, means that one contributes to achieving the other (see Map Key on page *xxii*). The occasional double-headed arrow implies mutual support.

One benchmark can contribute to another in different ways. For example, for students to grasp the principle that forces cause change in motion, logically they need some prior knowledge of what *forces* and *change in motion* are. But logic alone is seldom adequate to characterize growth of understanding. Psychological and developmental factors must also be considered. Often, to make progress toward understanding a benchmark, students will be aided by first understanding an earlier benchmark containing a less sophisticated version of the same concept.

Cognitive research specific to learning a particular idea can help identify important precursors, supporting connections from other subject areas, or common preconceptions. When specific research is unavailable, connections among benchmarks are guided by general principles of cognitive development, such as concrete-before-abstract and simple-before-complex.

For students to learn about the shape of the earth, for example, cognitive research reveals that it is important for them to have an early conception of the earth's gravity to account for why people on the "bottom" don't fall off. Ideas about the earth's gravity and its shape appear

Figure 3: Two connected benchmarks.

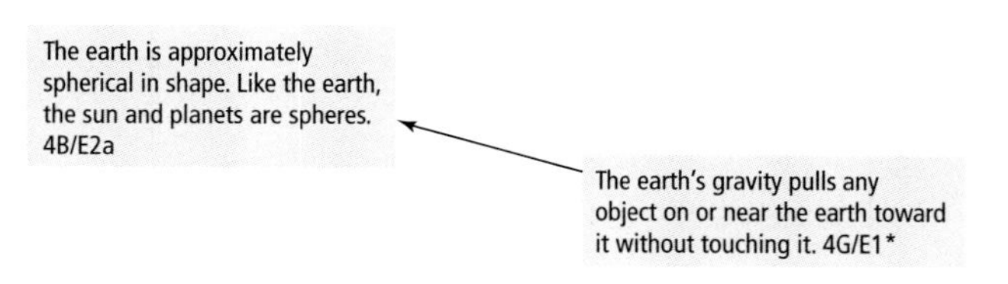

Figure 4: A strand of benchmarks about gravity.

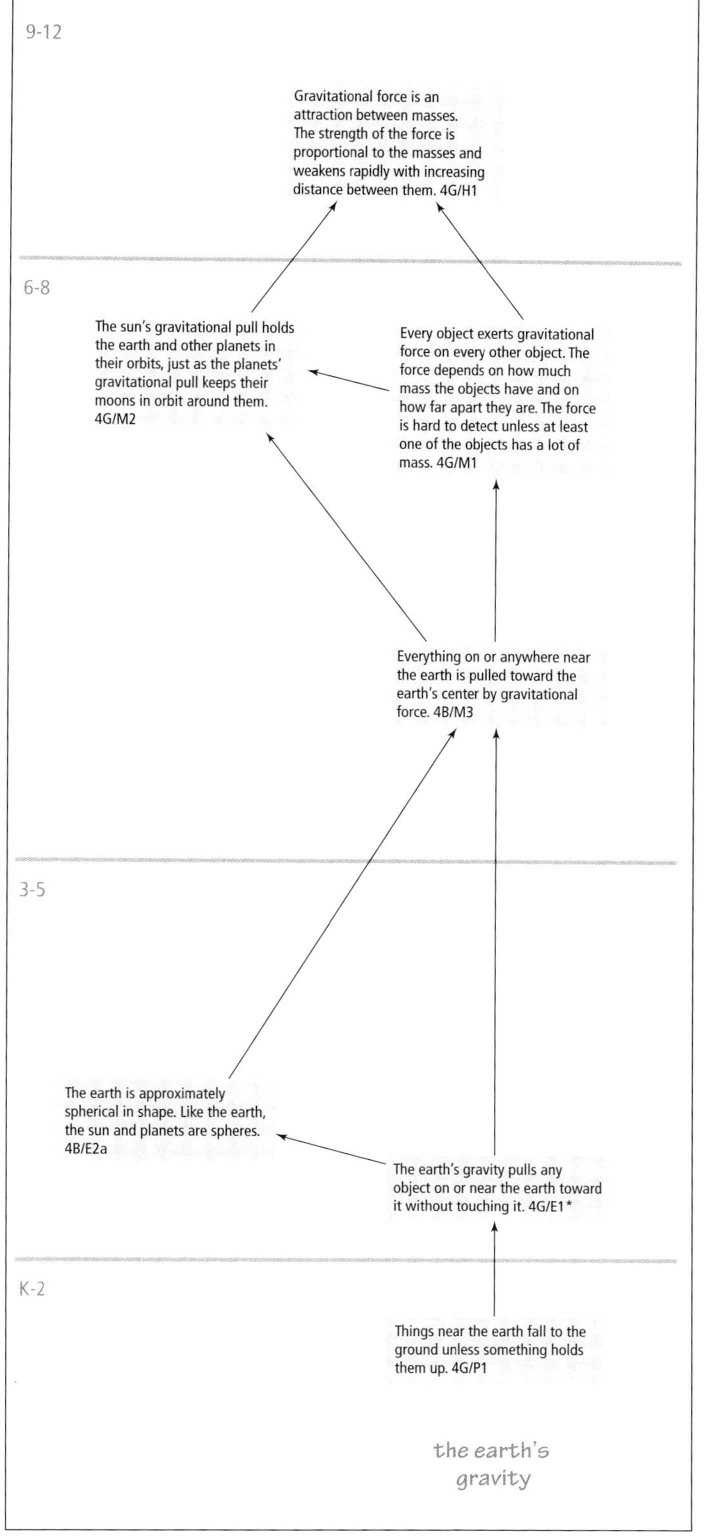


in two benchmarks in the 3-5 grade range. The arrow between the two boxes in Figure 3 implies that understanding one benchmark contributes to understanding the other.

The benchmark on the shape of the earth also plays a role in a strand of benchmarks about the earth's gravity, which leads to a benchmark about universal gravitation. Figure 4 shows the development of ideas about the earth's gravity in a strand of benchmarks that is part of the **GRAVITY** map published in Volume 1.

Achieving literacy in any particular topic involves the development of benchmarks at different levels of sophistication and often depends on benchmarks normally considered outside of the topic at hand.

## LABELED STRANDS

In each map, a few principal strands are pointed out to help the reader get a sense of its content (see Map Key on page *xxii*). The strands are meant to draw attention to some of the major concepts on the map and to how and when those concepts come together to form a coherent "story." These strands are labeled along the bottom of the map, and, where possible, relevant benchmarks are positioned in a column above each label. Not all strands have well-defined boundaries or even start at the bottom of the map. (See Figure 9 on page 119 for an example of how some strands overlap.) Nor are they the only strands possible for any particular map; strands are not meant to suggest the "correct" organization of benchmarks in a curriculum or prescribe a unique path of instruction.

## GRADE RANGES

The horizontal lines in each map delineate the grade ranges in which most students, given appropriate instruction, should be able to achieve particular benchmarks (see Map Key on page *xxii*). *Benchmarks* may be achieved in higher or lower grades depending on students' interests, abilities, and experience. But they ought not to be sorted into grades arbitrarily. When ideas and skills are pushed inappropriately lower in the grades—often to fit more into the curriculum—students' understanding suffers. When ideas and skills are left until too late, students' readiness and interest may wane.

A benchmark in a higher position within a grade range in a map is not necessarily more difficult than the other benchmarks in that same grade range. Design considerations often move benchmarks—up, down, or sideways—to allow connections to be drawn. What leads to what is indicated only by specific connections—arrows—between benchmarks.

## CONNECTIONS TO OTHER MAPS

We have not simply sorted the benchmarks from each section of *Benchmarks* into separate

Figure 5: The same benchmark in different maps.

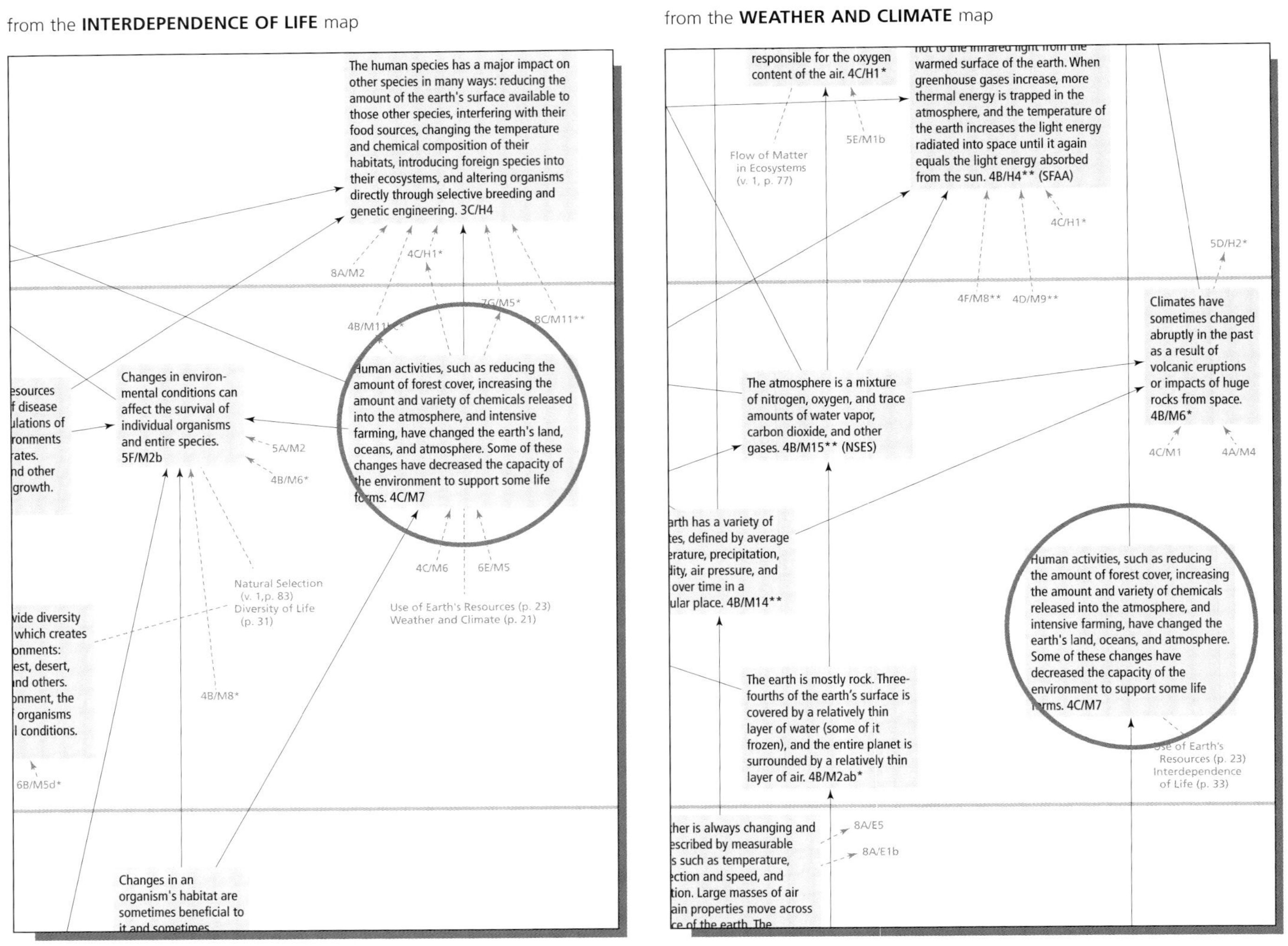


maps. Rather, each map includes benchmarks relevant to understanding a literacy topic, regardless of where else in *Atlas* those benchmarks appear. Indeed, all the benchmarks could be seen as part of one vast pattern of ideas, skills, and connections—a hypothetical "mother of all maps." The maps in this book are extracted from that grand fabric in pieces comprehensive enough to provide insight into a topic, but small enough to be manageable and useful.

Consequently, the same benchmarks—or even the same strands—can show up in more than one map. For example, a middle-school benchmark about energy can contribute to students learning about some very different topics in high school: states of matter, fusion in stars, and photosynthesis. If a benchmark plays an important role in several different maps, it may be included in each of them (likely showing somewhat different connections in each map). Because there may not be room on a map to display the full text of every benchmark that is linked to another, a benchmark code may be used instead, and a dotted arrow connects the code to related benchmarks.

Maps will not be understood as well, or used as effectively, if they are seen as independent entities. To help the reader keep in mind the notion of a larger fabric from which we have teased out this set of threads, the maps indicate connections between maps by cross-referencing benchmarks that appear in more than one map. As shown in Figure 5, a dotted line leads the reader from the benchmark to the titles and locations of the other maps in which that benchmark appears.

## READING STRAND MAPS

The strand maps are meant to be studied, wrestled with, and used. Although their complexity may seem intimidating at first, the effort put into studying them is worth it. In workshops and other presentations, we introduce maps by first showing only two or three benchmarks, taking time to let the participants investigate carefully the connections among the benchmarks and to think about what the connections mean. From there, it becomes easier to add benchmarks and gradually build toward the full map.

A similar incremental tack may help the reader approach and eventually get the most out of the maps in this book. Different approaches to a map may suit different uses, but most readers will find it helpful to focus first on a small part of the map. This could involve finding a central benchmark and exploring all its connections, or studying the labeled strands and following a sequence of connections across grade ranges.

## STRAND MAPS, CURRICULUM, AND INSTRUCTION

Strand maps do not prescribe any one curriculum design or sequence of instruction. Rather, they point out connections educators should *keep in mind* as they consider how to help students achieve science literacy. For example, the maps leave off many interesting relationships among benchmarks and topics that are not essential to the conceptual development of a topic but suggest a context in which the topic could be taught. A topic could be approached with an emphasis on some theme, such as models; from a particular historical perspective, such as the development of germ theory; or in a specific context, such as science and society. The maps leave those possibilities open rather than promote one approach out of many.

But even the connections that are shown explicitly on the maps do not prescribe curricular or instructional sequences. Connections among different disciplines that show up in the maps are not meant to suggest that the ideas or skills are learned best through integrated or cross-disciplinary courses. Salient connections could be addressed in a variety of course structures (including existing ones). There are many creative ways to bring benchmarks together, depending on circumstances unique to districts or individual classrooms (available resources, class interest in a particular project, and so on).

Maps do not predict or dictate the interplay of ideas and skills that can occur in different settings. However, they do provide a useful starting point for all those who study or make decisions about the design and implementation of curriculum and instruction. It is our hope that research on student learning and other relevant investigations will eventually lead to maps that are built on empirically tested assumptions. Toward that end, we invite readers to share with us their experiences and research results that make use of *Atlas* maps.

# MAP KEY

**BENCHMARKS**
are specific learning goals derived mostly from *Benchmarks for Science Literacy* but also from *Science for All Americans* and *National Science Education Standards*. Colored boxes indicate knowledge goals; bordered boxes indicate skill goals. Some benchmarks have been split into two or more ideas which appear in separate boxes. (See page *xvi.*)

**CONNECTING ARROWS**
indicate that achieving one benchmark contributes to achieving the other. The exact meaning of a connection is not indicated explicitly, but connections can be based on the logic of the subject matter or on cognitive research about how students learn. (See page *xviii.*)

**GRADE RANGES**
suggest when most students could achieve these benchmarks. A benchmark's position within a grade range does not indicate the grade in which it should be taught, nor does its position indicate that it should be taught before or after another benchmark unless there is an arrow connecting them. (See page *xix.*)

**BENCHMARK CODES**
indicate chapter, section, grade range, and number of the corresponding goal statements in *Benchmarks for Science Literacy*. Letters, asterisks, and acronyms following the code provide additional information about the benchmark. (See page *xvii.*)

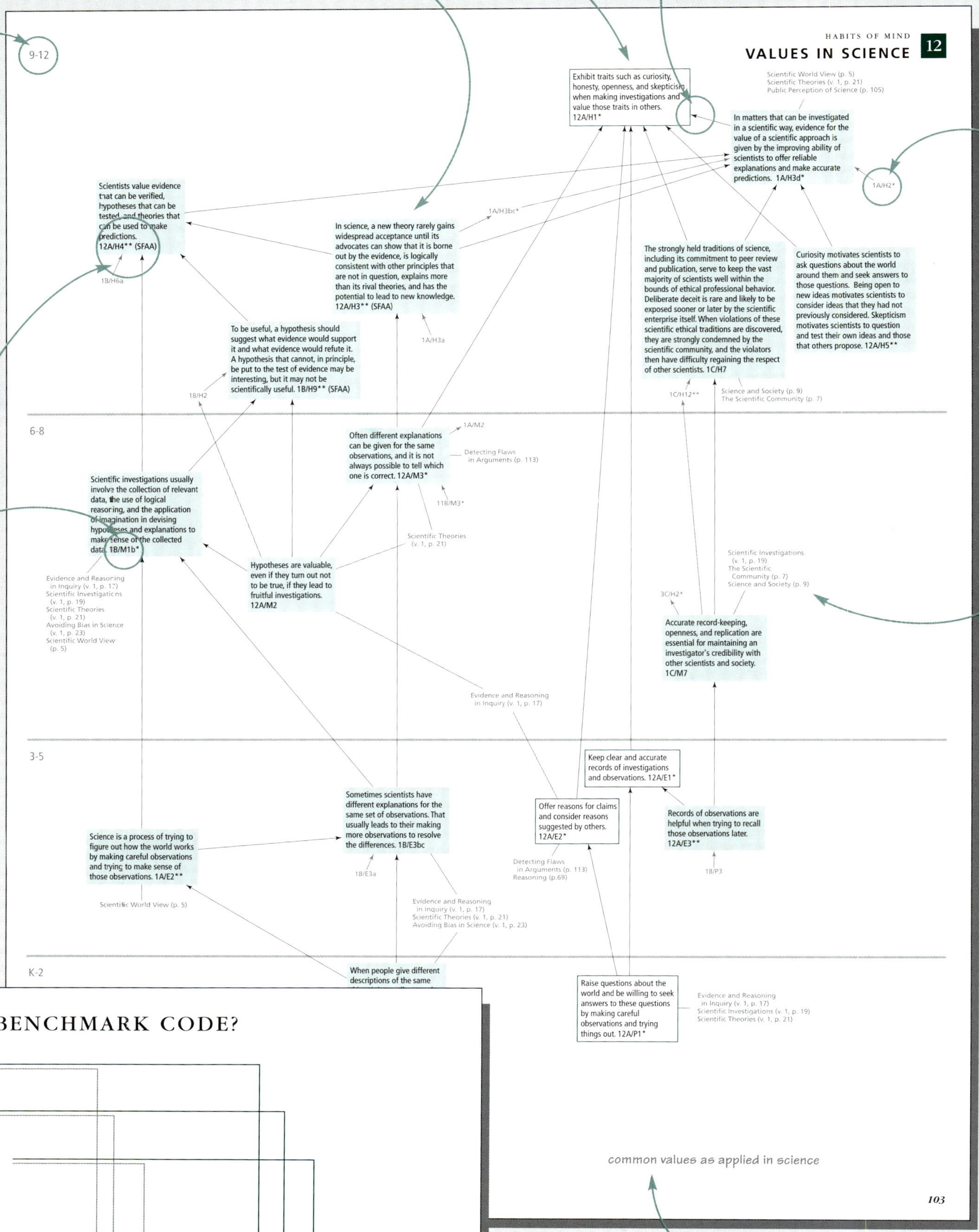


**OFF-MAP CONNECTIONS**
show links to the codes of closely related benchmarks when it is not possible to include the full text of the benchmark on a map. Arrows in off-map connections imply the same relationship between benchmarks as they do when they connect boxes to boxes. Use the Index of Mapped Benchmarks on page 124 to search by benchmark code for maps on which a benchmark appears in its full text. (See page *xx.*)

**CROSS-REFERENCES TO OTHER MAPS**
indicate that the benchmark also appears on the maps that are listed. (See page *xx.*)

**STRAND LABELS**
help the reader find things in the map and get a sense of the map's content. Strands loosely suggest ideas or skills that develop over time. Strands often interweave and share benchmarks. (See page *xix.*)

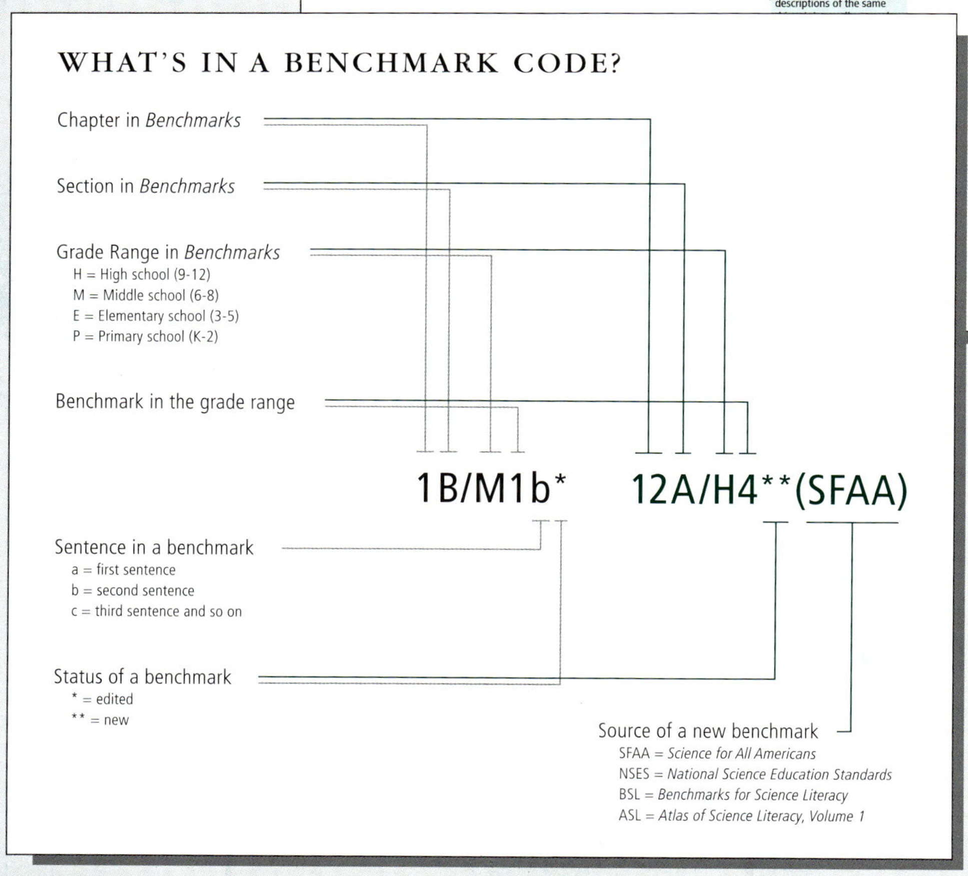

# THE STRAND MAPS

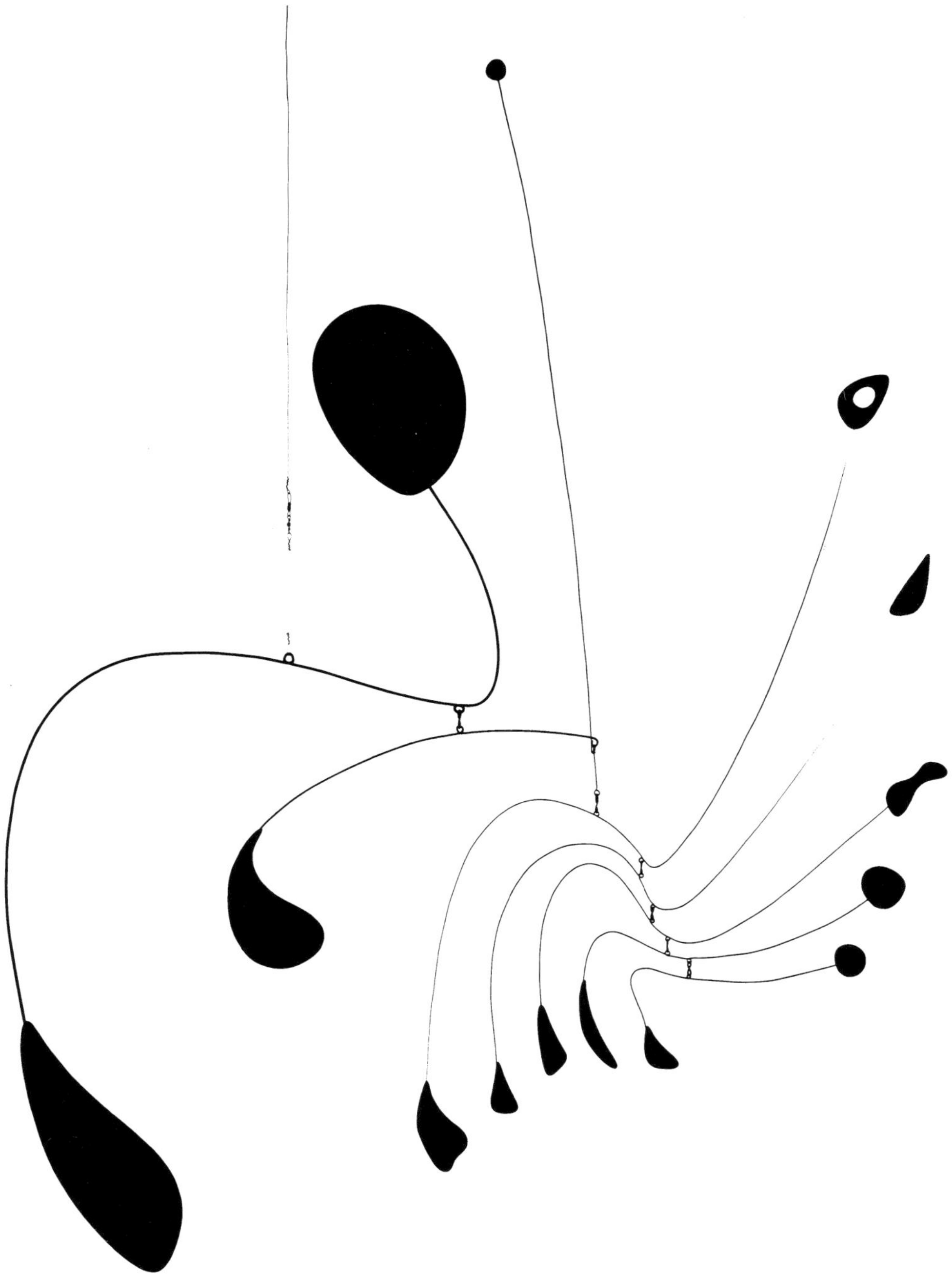

Alexander Calder *Hanging Spider,* 1940

*Over the course of human history, people have developed many interconnected and validated ideas about the physical, biological, psychological, and social worlds. Those ideas have enabled successive generations to achieve an increasingly comprehensive and reliable understanding of the human species and its environment. The means used to develop these ideas are particular ways of observing, thinking, experimenting, and validating. These ways represent a fundamental aspect of the nature of science and reflect how science tends to differ from other modes of knowing.*

SCIENCE FOR ALL AMERICANS

# CHAPTER 1



# THE NATURE OF SCIENCE

The nature of science involves the basic values and beliefs that make up the scientific world view, the way scientists go about their work, and the general culture of the scientific enterprise. When people know how scientists go about their work and reach scientific conclusions, and what the limitations of such conclusions are, people are more likely to react thoughtfully to scientific claims and less likely to reject them out of hand or accept them uncritically. Once people gain a good sense of how science operates—along with a basic inventory of key science concepts as a basis for learning more later—they can follow the science adventure story as it plays out during their lifetimes.

The list on the left of this page shows the maps for this chapter and indicates the volume of *Atlas of Science Literacy* in which each appears. Those in green typeface are in *Volume 2* and those in gray are in *Volume 1*.

# SCIENTIFIC WORLD VIEW (1A)

Scientists share certain basic assumptions about the nature of the world and what can be learned about it: The world can be understood through careful, systematic study; scientific knowledge produced through such a process is both durable and subject to change; the scientific process cannot answer some questions, such as those about values and beliefs. In order to follow the science story as it unfolds, students need to understand these shared assumptions.

The map is organized around three strands—*limits of science, investigating a knowable world,* and *continuity and change*. In the elementary grades, the focus is on what can be learned from observation and experimentation. In middle school, ideas are introduced about the modifiability of science that results from new discoveries and about what cannot be studied in a scientific way. In high school, various historical episodes serve as examples of generalizations about continuity and change in science and the assumptions underlying a scientific world view.

Several historical episodes mapped in Chapter 10: HISTORICAL PERSPECTIVES illustrate how scientific knowledge is judged, modified, and replaced and exemplify major shifts in the scientific view of how the world works. The **CLASSICAL MECHANICS** map illustrates scientists' assumptions about the unity and understandability of the natural world.

### NOTES

The 9-12 benchmark "Science is based on..." in the *investigating a knowable world* strand focuses on two related but separate premises. One is that by careful systematic study, people can figure out how the world works. The other is that the universe is a unified system and knowledge gained from studying one part of it can be applied to other parts.

The left-hand side of the strand includes ideas about the unity of nature. Nearly all of these benchmarks also appear in the **SCIENTIFIC INVESTIGATIONS** map in *Atlas 1* where they are part of a sequence of ideas about the importance of reliability in investigations. The right-hand side of the strand unpacks what is meant by careful systematic study of the world. In the grades 3-5 range, a new benchmark (1A/E2) describes what science is and its purpose at a level of complexity that most early elementary students can understand.

Benchmarks in the *continuity and change* strand also play a role in the *theory modification* strand in the **SCIENTIFIC THEORIES** map in *Atlas 1*.

### RESEARCH IN *BENCHMARKS*

Although most students believe that scientific knowledge changes, they typically think changes occur mainly in facts and mostly through the invention of improved technology for observation and measurement. They do not recognize that changed theories sometimes suggest new observations or reinterpretation of previous observations (Aikenhead, 1987; Lederman & O'Malley, 1990; Waterman, 1983). Some research indicates that it is difficult for middle-school students to understand the development of scientific knowledge through the interaction of theory and observation (Carey et al., 1989), but the lack of long-term teaching interventions to investigate this issue makes it difficult to conclude that students can or cannot gain that understanding at this grade level.

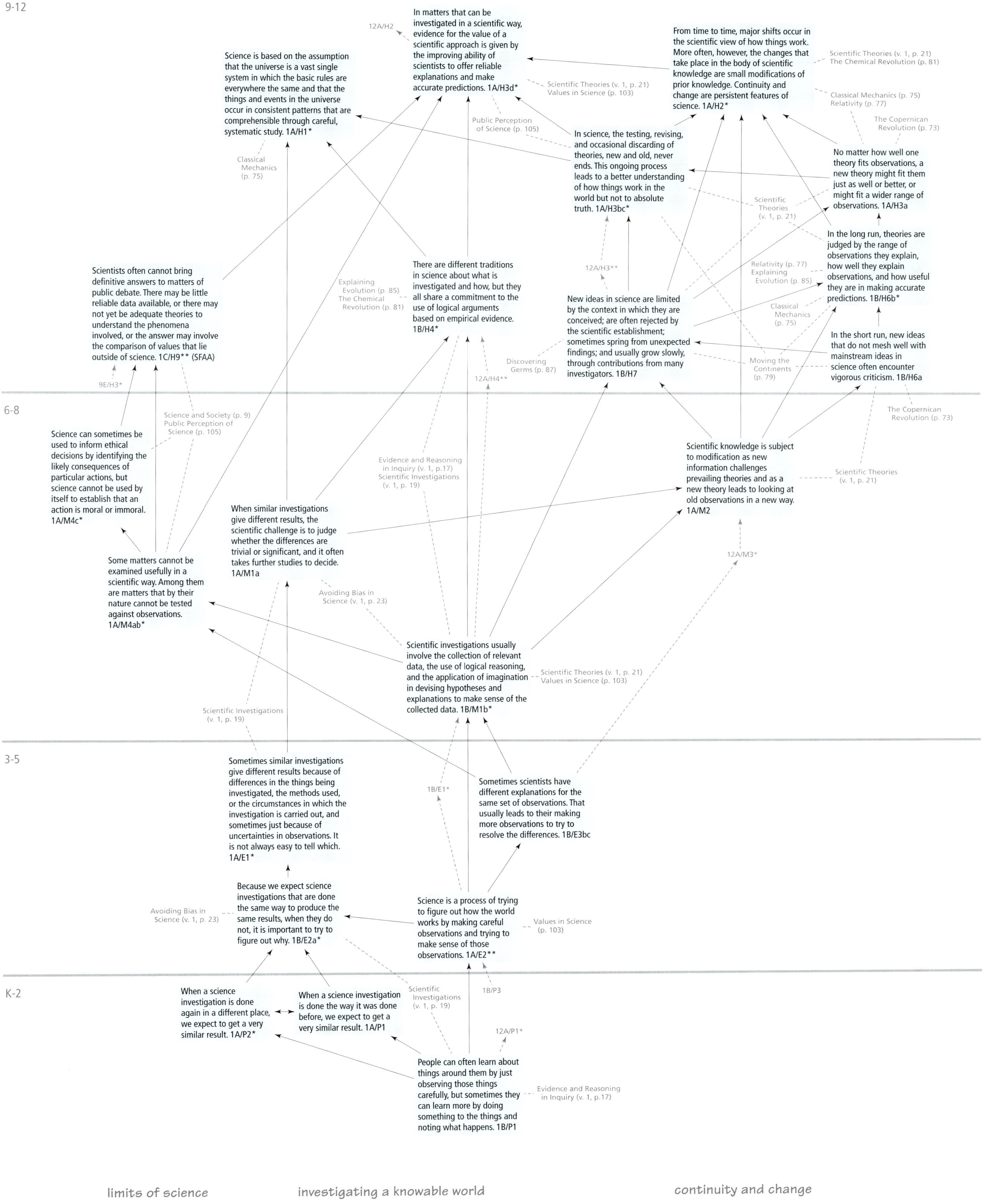
9-12
In matters that can be investigated in a scientific way, evidence for the value of a scientific approach is given by the improving ability of scientists to offer reliable explanations and make accurate predictions. 1A/H3d*
12A/H2
Science is based on the assumption that the universe is a vast single system in which the basic rules are everywhere the same and that the things and events in the universe occur in consistent patterns that are comprehensible through careful, systematic study. 1A/H1*
From time to time, major shifts occur in the scientific view of how things work. More often, however, the changes that take place in the body of scientific knowledge are small modifications of prior knowledge. Continuity and change are persistent features of science. 1A/H2*
Scientific Theories (v. 1, p. 21)
The Chemical Revolution (p. 81)
Scientific Theories (v. 1, p. 21)
Values in Science (p. 103)
Classical Mechanics (p. 75)
Relativity (p. 77)
The Copernican Revolution (p. 73)
Public Perception of Science (p. 105)
In science, the testing, revising, and occasional discarding of theories, new and old, never ends. This ongoing process leads to a better understanding of how things work in the world but not to absolute truth. 1A/H3bc*
No matter how well one theory fits observations, a new theory might fit them just as well or better, or might fit a wider range of observations. 1A/H3a
Classical Mechanics (p. 75)
Scientific Theories (v. 1, p. 21)
In the long run, theories are judged by the range of observations they explain, how well they explain observations, and how useful they are in making accurate predictions. 1B/H6b*
Scientists often cannot bring definitive answers to matters of public debate. There may be little reliable data available, or there may not yet be adequate theories to understand the phenomena involved, or the answer may involve the comparison of values that lie outside of science. 1C/H9** (SFAA)
Explaining Evolution (p. 85)
The Chemical Revolution (p. 81)
There are different traditions in science about what is investigated and how, but they all share a commitment to the use of logical arguments based on empirical evidence. 1B/H4*
12A/H3**
Relativity (p. 77)
Explaining Evolution (p. 85)
New ideas in science are limited by the context in which they are conceived; are often rejected by the scientific establishment; sometimes spring from unexpected findings; and usually grow slowly, through contributions from many investigators. 1B/H7
Classical Mechanics (p. 75)
In the short run, new ideas that do not mesh well with mainstream ideas in science often encounter vigorous criticism. 1B/H6a
Discovering Germs (p. 87)
Moving the Continents (p. 79)
12A/H4**
9E/H3*
The Copernican Revolution (p. 73)
6-8
Science and Society (p. 9)
Public Perception of Science (p. 105)
Science can sometimes be used to inform ethical decisions by identifying the likely consequences of particular actions, but science cannot be used by itself to establish that an action is moral or immoral. 1A/M4c*
Evidence and Reasoning in Inquiry (v. 1, p.17)
Scientific Investigations (v. 1, p. 19)
Scientific knowledge is subject to modification as new information challenges prevailing theories and as a new theory leads to looking at old observations in a new way. 1A/M2
Scientific Theories (v. 1, p. 21)
When similar investigations give different results, the scientific challenge is to judge whether the differences are trivial or significant, and it often takes further studies to decide. 1A/M1a
Some matters cannot be examined usefully in a scientific way. Among them are matters that by their nature cannot be tested against observations. 1A/M4ab*
12A/M3*
Avoiding Bias in Science (v. 1, p. 23)
Scientific investigations usually involve the collection of relevant data, the use of logical reasoning, and the application of imagination in devising hypotheses and explanations to make sense of the collected data. 1B/M1b*
Scientific Theories (v. 1, p. 21)
Values in Science (p. 103)
Scientific Investigations (v. 1, p. 19)
3-5
Sometimes similar investigations give different results because of differences in the things being investigated, the methods used, or the circumstances in which the investigation is carried out, and sometimes just because of uncertainties in observations. It is not always easy to tell which. 1A/E1*
1B/E1*
Sometimes scientists have different explanations for the same set of observations. That usually leads to their making more observations to try to resolve the differences. 1B/E3bc
Avoiding Bias in Science (v. 1, p. 23)
Because we expect science investigations that are done the same way to produce the same results, when they do not, it is important to try to figure out why. 1B/E2a*
Science is a process of trying to figure out how the world works by making careful observations and trying to make sense of those observations. 1A/E2**
Values in Science (p. 103)
K-2
When a science investigation is done again in a different place, we expect to get a very similar result. 1A/P2*
When a science investigation is done the way it was done before, we expect to get a very similar result. 1A/P1
Scientific Investigations (v. 1, p. 19)
1B/P3
12A/P1*
People can often learn about things around them by just observing those things carefully, but sometimes they can learn more by doing something to the things and noting what happens. 1B/P1
Evidence and Reasoning in Inquiry (v. 1, p.17)
limits of science
investigating a knowable world
continuity and change

# THE SCIENTIFIC COMMUNITY (1C)

The scientific community plays a significant role in contemporary life, informing invention and public policy and employing a large and diverse number of people around the world. Understanding how scientific activity is organized and funded, and how its findings are communicated, gives students an appreciation of scientific inquiry as a complex social activity and of ways in which they can participate—whether as scientists or as citizens.

The map is organized around four strands—*disciplines, employment and funding, people,* and *scientific communication.* Earlier grades emphasize the inclusiveness of the scientific endeavor and the importance of clear and accurate communication. Later grades emphasize the blurred boundaries between scientific disciplines, the role of society in setting funding priorities, and the importance of communication in both disseminating scientific findings and in making findings available for scrutiny.

**NOTES**

Several benchmarks in the *employment and funding* strand also appear in the **SCIENCE AND SOCIETY** map and deal with influences from within the scientific community itself and from society in general that affect the direction and priorities of funding for scientific research.

The *people* strand focuses on the characteristics of the scientific work force and on inequities in access to and participation in the scientific enterprise. The double-headed arrow between the 6-8 benchmarks "Important contributions..." and "Until recently, women and racial minorities..." illustrates the close relationship between these two ideas and indicates that each benchmark helps clarify the other.

As an illustration of ideas in the *scientific communication* strand, the story of Lavoisier's development of a system for naming substances and describing reactions in the Chapter 10 map, **THE CHEMICAL REVOLUTION,** provides an example of the role of clear communication in advancing science.

**RESEARCH IN *BENCHMARKS***

When asked to describe their views about science in general, high-school students portray scientists as brilliant, dedicated, and essential to the world. However, when asked about science as a career, students respond with a negative image of scientific work and scientists. They see scientific work as dull and rarely rewarding, and scientists as bearded, balding, working alone in the laboratory, isolated, and lonely (Mead & Metraux, 1957). This image of scientists has also been frequently documented among elementary- and middle-school students (Fort & Varney, 1989; Newton & Newton, 1992). Some research suggests that this image may represent students' knowledge of the public stereotype rather than their personal views and knowledge of science and scientists (Boylan, Hill, Wallace, & Wheeler, 1992).

Some students of all ages believe science mainly invents things or solves practical problems rather than exploring and understanding the world. Some high-school students believe that moral values and personal motives do not influence a scientist's contributions to the public debate about science and technology and think that scientists are more capable than others to decide those issues (Aikenhead, 1987; Fleming, 1986a, 1986b, 1987).

"THAT'S IT? THAT'S PEER REVIEW?"

## 9-12

The direction of scientific research is affected by informal influences within the culture of science itself, such as prevailing opinion on which questions are most interesting or which methods of investigation are most likely to be fruitful. Elaborate processes involving scientists themselves have been developed to decide which research proposals receive funding, and committees of scientists regularly review progress in various disciplines to recommend general priorities for funding. 1C/H11** (SFAA)

Avoiding Bias in Science (v. 1, p. 23)

1C/H3a

1C/H6d*

The strongly held traditions of science, including its commitment to peer review and publication, serve to keep the vast majority of scientists well within the bounds of ethical professional behavior. Deliberate deceit is rare and likely to be exposed sooner or later by the scientific enterprise itself. When violations of these scientific ethical traditions are discovered, they are strongly condemned by the scientific community, and the violators then have difficulty regaining the respect of other scientists. 1C/H7

Science and Society (p. 9)
Values in Science (p. 103)

1A/H1*

Public Perception of Science (p. 105)

Science disciplines differ from one another in what is studied, techniques used, and outcomes sought, but they share a common purpose and philosophy, and all are part of the same scientific enterprise. Although each discipline provides a conceptual structure for organizing and pursuing knowledge, many problems are studied by scientists using information and skills from many disciplines. Disciplines do not have fixed boundaries, and it happens that new scientific disciplines are being formed where existing ones meet and that some subdisciplines spin off to become new disciplines in their own right. 1C/H4

Funding influences the direction of science by virtue of the decisions that are made on which research to support. Research funding comes from various federal government agencies, industry, and private foundations. 1C/H8

Splitting the Atom (p.83)

Science and Society (p. 9)

Because science is a human activity, what is valued in society influences what is valued in science. 1C/H10** (SFAA)

Public Perception of Science (p. 105)

Scientists' nationality, sex, ethnic origin, age, political convictions, and so on may incline them to look for or emphasize one or another kind of evidence or interpretation. 1B/H8** (SFAA)

The early Egyptian, Greek, Chinese, Hindu, and Arabic cultures are responsible for many scientific and mathematical ideas and technological inventions. Modern science is based on traditions of thought that came together in Europe about 500 years ago. People from all cultures now contribute to that tradition. 1C/H1*

3C/H2*

10F/H2*

The dissemination of scientific information is crucial to its progress. Some scientists present their findings and theories in papers that are delivered at meetings or published in scientific journals. Those papers enable scientists to inform others about their work, to expose their ideas to criticism by other scientists, and, of course, to stay abreast of scientific developments around the world. 1C/H12** (SFAA)

1A/H2*

10H/H4*

10D/H3*

10A/H4*

## 6-8

Scientists' personal interests and viewpoints can influence the questions they investigate. 1C/M8** (SFAA)

No matter who does science and mathematics or invents things, or when or where they do it, the knowledge and technology that result can eventually become available to everyone in the world. 1C/M3

1A/M2

1A/M3

Until recently, women and racial minorities, because of restrictions on their education and employment opportunities, were essentially left out of much of the formal work of the science establishment; the remarkable few who overcame those obstacles were even then likely to have their work disregarded by the science establishment. 1C/M2

Science and Society (p. 9)

Important contributions to the advancement of science, mathematics, and technology have been made by different kinds of people, in different cultures, at different times. 1C/M1

1C/M6

Accurate recordkeeping, openness, and replication are essential for maintaining an investigator's credibility with other scientists and society. 1C/M7

Scientists are employed by colleges and universities, business and industry, hospitals, and many government agencies. Their places of work include offices, classrooms, laboratories, farms, factories, and natural field settings ranging from space to the ocean floor. 1C/M4

Scientists are linked to other scientists worldwide both personally and through international scientific organizations. 1C/M9** (BSL)

Scientists differ greatly in what phenomena they study and how they go about their work. 1B/M1a

1B/E4

Science and Society (p. 9)
Scientific Investigations (v. 1, p. 19)
Values in Science (p. 103)

## 3-5

Doing science involves many different kinds of work and engages men and women of all ages and backgrounds. 1C/E3

Science is an adventure that people everywhere can take part in, as they have for many centuries. 1C/E1

Clear communication is an essential part of doing science. It enables scientists to inform others about their work, expose their ideas to criticism by other scientists, and stay informed about scientific discoveries around the world. 1C/E2

Scientific Investigations (v. 1, p. 19)
Avoiding Bias in Science (v. 1, p. 23)

Human beings have different interests, motivations, skills, and talents. 6D/E1

## K-2

6D/P3

7B/E1*

1B/P4

Describing things as accurately as possible is important in science because it enables people to compare their observations with those of others. 1B/P3

In doing science, it is often helpful to work with a team and to share findings with others. All team members should reach their own individual conclusions, however, about what the findings mean. 1C/P2

Everybody can do science and invent things and ideas. 1C/P1

Evidence and Reasoning in Inquiry (v. 1, p. 17)
Scientific Investigations (v. 1, p. 19)

Avoiding Bias in Science (v. 1, p. 23)

disciplines | employment & funding | people | scientific communication

# SCIENCE AND SOCIETY (1C)

Science both affects and is affected by society. Scientists influence social decision-making; new ideas in science challenge our views of the world, and new applications extend our abilities to shape it. At the same time, social and economic forces influence what research will be undertaken, paid attention to, and applied. Ethical principles inform the conduct of science and serve to keep the vast majority of scientists well within the bounds of ethical professional behavior.

The map is organized around three strands—*ethics in research, society affects science,* and *science affects society.* There are only a few benchmarks before the 6-8 grade range, in part because students need a clear idea of what science is as an intellectual endeavor before they can begin exploring how it affects and is affected by society.

Related maps on **EVIDENCE AND REASONING IN INQUIRY** in *Atlas 1* and **DETECTING FLAWS IN ARGUMENTS** describe the principles of scientific inquiry and reasoning that can help students understand the role that scientists play in public affairs. Real-world examples of how science and society interact are presented in many of the Chapter 10 maps, which outline important episodes in the history of science, such as **SPLITTING THE ATOM, DISCOVERING GERMS**, and **THE INDUSTRIAL REVOLUTION**. Examples that involve technology are explored in the **TECHNOLOGY AND SCIENCE** map in this volume and the **INTERACTION OF TECHNOLOGY AND SOCIETY** map in *Atlas 1*.

### NOTES

Three new 9-12 benchmarks in the *society affects science* strand relate to the idea that scientists are members of society and science and bring their own cultural backgrounds and views to both realms. The benchmarks have been added to clarify further the social nature of science ("Scientists' nationality, sex, ethnic origin..." and "Because science is a human activity...") and the limits of scientific expertise in resolving public issues ("Scientists often cannot bring definitive answers..."). Connections between benchmarks in the *science affects society* strand and the *ethics in research* strand center on the need for scientists to maintain credibility within society.

### RESEARCH IN *BENCHMARKS*

Some students of all ages believe science mainly invents things or solves practical problems rather than exploring and understanding the world. Some high-school students believe that moral values and personal motives do not influence a scientist's contributions to the public debate about science and technology and think that scientists are more capable than others to decide those issues (Aikenhead, 1987; Fleming 1986a, 1986b, 1987).

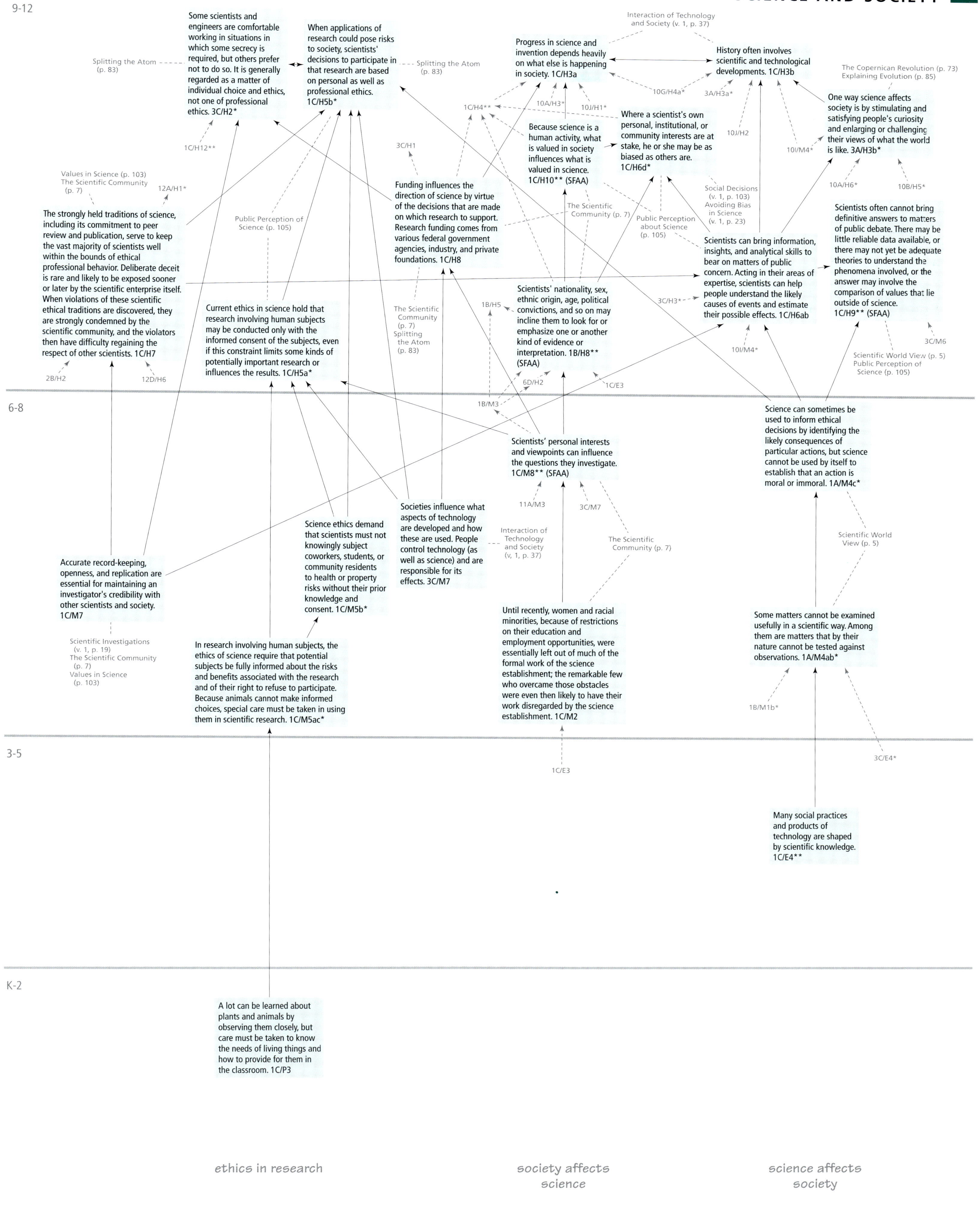
9-12
6-8
3-5
K-2
Some scientists and engineers are comfortable working in situations in which some secrecy is required, but others prefer not to do so. It is generally regarded as a matter of individual choice and ethics, not one of professional ethics. 3C/H2*
Splitting the Atom (p. 83)
When applications of research could pose risks to society, scientists' decisions to participate in that research are based on personal as well as professional ethics. 1C/H5b*
Splitting the Atom (p. 83)
Interaction of Technology and Society (v. 1, p. 37)
Progress in science and invention depends heavily on what else is happening in society. 1C/H3a
History often involves scientific and technological developments. 1C/H3b
The Copernican Revolution (p. 73)
Explaining Evolution (p. 85)
One way science affects society is by stimulating and satisfying people's curiosity and enlarging or challenging their views of what the world is like. 3A/H3b*
10G/H4a*
3A/H3a*
10A/H3*
10J/H1*
1C/H4**
Because science is a human activity, what is valued in society influences what is valued in science. 1C/H10** (SFAA)
Where a scientist's own personal, institutional, or community interests are at stake, he or she may be as biased as others are. 1C/H6d*
10J/H2
10I/M4*
1C/H12**
3C/H1
Values in Science (p. 103)
The Scientific Community (p. 7)
12A/H1*
The strongly held traditions of science, including its commitment to peer review and publication, serve to keep the vast majority of scientists well within the bounds of ethical professional behavior. Deliberate deceit is rare and likely to be exposed sooner or later by the scientific enterprise itself. When violations of these scientific ethical traditions are discovered, they are strongly condemned by the scientific community, and the violators then have difficulty regaining the respect of other scientists. 1C/H7
Public Perception of Science (p. 105)
Funding influences the direction of science by virtue of the decisions that are made on which research to support. Research funding comes from various federal government agencies, industry, and private foundations. 1C/H8
The Scientific Community (p. 7)
Public Perception about Science (p. 105)
Social Decisions (v. 1, p. 103)
Avoiding Bias in Science (v. 1, p. 23)
10A/H6*
10B/H5*
Scientists often cannot bring definitive answers to matters of public debate. There may be little reliable data available, or there may not yet be adequate theories to understand the phenomena involved, or the answer may involve the comparison of values that lie outside of science. 1C/H9** (SFAA)
Scientists can bring information, insights, and analytical skills to bear on matters of public concern. Acting in their areas of expertise, scientists can help people understand the likely causes of events and estimate their possible effects. 1C/H6ab
3C/H3*
Current ethics in science hold that research involving human subjects may be conducted only with the informed consent of the subjects, even if this constraint limits some kinds of potentially important research or influences the results. 1C/H5a*
The Scientific Community (p. 7)
Splitting the Atom (p. 83)
1B/H5
Scientists' nationality, sex, ethnic origin, age, political convictions, and so on may incline them to look for or emphasize one or another kind of evidence or interpretation. 1B/H8** (SFAA)
10I/M4*
3C/M6
Scientific World View (p. 5)
Public Perception of Science (p. 105)
2B/H2
12D/H6
6D/H2
1C/E3
1B/M3
Scientists' personal interests and viewpoints can influence the questions they investigate. 1C/M8** (SFAA)
Science can sometimes be used to inform ethical decisions by identifying the likely consequences of particular actions, but science cannot be used by itself to establish that an action is moral or immoral. 1A/M4c*
11A/M3
3C/M7
Societies influence what aspects of technology are developed and how these are used. People control technology (as well as science) and are responsible for its effects. 3C/M7
Science ethics demand that scientists must not knowingly subject coworkers, students, or community residents to health or property risks without their prior knowledge and consent. 1C/M5b*
Interaction of Technology and Society (v, 1, p. 37)
The Scientific Community (p. 7)
Scientific World View (p. 5)
Accurate record-keeping, openness, and replication are essential for maintaining an investigator's credibility with other scientists and society. 1C/M7
Scientific Investigations (v. 1, p. 19)
The Scientific Community (p. 7)
Values in Science (p. 103)
In research involving human subjects, the ethics of science require that potential subjects be fully informed about the risks and benefits associated with the research and of their right to refuse to participate. Because animals cannot make informed choices, special care must be taken in using them in scientific research. 1C/M5ac*
Until recently, women and racial minorities, because of restrictions on their education and employment opportunities, were essentially left out of much of the formal work of the science establishment; the remarkable few who overcame those obstacles were even then likely to have their work disregarded by the science establishment. 1C/M2
Some matters cannot be examined usefully in a scientific way. Among them are matters that by their nature cannot be tested against observations. 1A/M4ab*
1B/M1b*
3C/E4*
1C/E3
Many social practices and products of technology are shaped by scientific knowledge. 1C/E4**
A lot can be learned about plants and animals by observing them closely, but care must be taken to know the needs of living things and how to provide for them in the classroom. 1C/P3
ethics in research
society affects science
science affects society

C. D. Doxiadias *Model of boulevard network of central Paris,* 20th century

*Mathematics relies on both logic and creativity, and it is pursued both for a variety of practical purposes and for its intrinsic interest. For some people, and not only professional mathematicians, the essence of mathematics lies in its beauty and its intellectual challenge. For others, including many scientists and engineers, the chief value of mathematics is how it applies to their own work. Because mathematics plays such a central role in modern culture, some basic understanding of the nature of mathematics is requisite for scientific literacy. To achieve this, students need to perceive mathematics as part of the scientific endeavor, comprehend the nature of mathematical thinking, and become familiar with key mathematical ideas and skills.*

SCIENCE FOR ALL AMERICANS

CHAPTER 2

# THE NATURE OF MATHEMATICS



As the science of patterns and relationships, mathematics shares many of the features of other sciences. Particularly relevant similarities are the belief in underlying order, the ideals of honesty and openness in reporting research, the importance of criticism by colleagues in judging the value of new work, and the essential role played by imagination. In *Everybody Counts* (1989), a report on mathematics education published in the same year as *Science for All Americans*, the National Research Council describes the relationship of mathematics to science and technology this way:

> …mathematics today is a diverse discipline that deals with data, measurements, and observations from science; with inference, deduction, and proof; and with mathematical models of natural phenomena, of human behavior, and of social systems…. Experience with mathematical modes of thought builds mathematical power—a capacity of mind of increasing value in this technological age that enables one to read critically, to identify fallacies, to detect bias, to assess risk, and to suggest alternatives. Mathematics empowers us to understand better the information-laden world in which we live.

The list on the left of this page shows the maps for this chapter and indicates the volume of *Atlas of Science Literacy* in which each appears. Those in green typeface are in *Volume 2* and those in gray are in *Volume 1*.

# NATURE OF MATHEMATICS (2AB)

The universe is made up of galaxies, mountains, organisms, vehicles, and a wide variety of other things, each seemingly unique. Moreover, these things interact with one another in all sorts of ways, often violently but sometimes with great subtlety. But thanks to mathematics, people are able to think about the world of objects and events and to communicate those thoughts in ways that reveal unity and order. Mathematics is a universal language for describing patterns and relationships, both about abstractions and about objects and events in the real world. The results of theoretical and applied mathematics often influence one another, contributing to a better understanding of the world.

The map is organized around three strands—*study of patterns, universal language for describing things,* and *connections to science and technology*. They reflect key understandings students are expected to develop about mathematics as the study of patterns that involves the interplay of theory and application and that both draws on and contributes to science and technology.

In the elementary grades, the focus is on the use of quantities and shapes to describe real-world objects and events. Middle-school benchmarks emphasize the role of mathematics in representing change over time, negative quantities, and data sets. High-school benchmarks emphasize the role of mathematics in modeling real-world events and technological designs. Connections are made to benchmarks describing the contribution of computers to these mathematical activities.

In addition to other maps in this chapter, the map draws upon and contributes to maps in Chapter 9: THE MATHEMATICAL WORLD, in Chapter 11: COMMON THEMES, and in Chapter 12: HABITS OF MIND.

The use of mathematical models is a special case of how mathematics contributes to science and technology. The progression of understanding about mathematical models is laid out in the **MATHEMATICAL MODELS** map in *Atlas 1*. The use of mathematical models in science and technology is developed in greater depth in the **MODELS** map.

## NOTES

Many specific ideas about the uses of mathematics (such as using shapes and numbers to describe objects and events or using graphs to help make predictions about phenomena), contribute to the 6-8 benchmark "Mathematics is helpful in almost every kind of human endeavor...." More abstract ideas about and uses of mathematics (such as those involving negative numbers, measures of central tendency, and mathematical modeling) contribute to the 9-12 benchmark "Mathematics provides a precise language..." and to a more in-depth understanding of the utility of mathematics in all fields of human endeavor, not just in science and technology.

Most mathematical theories have applications, and much work in mathematics is motivated by seeking answers to applied problems, often in science and technology. This connection is depicted by the arrows linking 9-12 benchmarks "Developments in mathematics often stimulate..." and "Developments in science or technology often stimulate..." to the 9-12 benchmark "Theories and applications in mathematical work...."

"THIS IS THE PART I ALWAYS HATE."

## RESEARCH IN *BENCHMARKS*

Early reviews of the research (e.g., McLeod, 1992; Schoenfeld, 1992; Thompson, 1992) support continuing studies of beliefs, attitudes, and values about the nature of mathematics. For descriptions of international research programs, see the volume edited by Leder, Pehkonen, and Toerner (2002), and the special issue on affect published by *Educational Studies in Mathematics* (e.g., Zan, Brown, Evans, & Hannula, 2006). For a more comprehensive review of recent literature, with an emphasis on teacher beliefs, see Philipp (in press). Data on student beliefs and attitudes are also a component of large-scale studies, such as the National Assessment of Educational Progress and the Third International Mathematics and Science Study (see, e.g., Silver & Kenney, 2000). Investigations of students' beliefs about the nature of mathematics are illuminated by Schoenfeld (1992) and summarized in a report from the National Research Council (Kilpatrick, Swafford, & Findell, 2001). These studies examined difficulties arising from students' beliefs about the nature of mathematical problem solving and from their perceptions of the role of memorization in learning mathematics and of mathematics as rule-oriented versus process-oriented or as a static versus a dynamic discipline. There has been less emphasis on students' understanding of the relationships between mathematics, science, and technology or to the nature of mathematical inquiry as a modeling process.

Research on student understanding of early algebra ideas, such as patterns and relationships, has received increased attention. Early studies suggested students have difficulty connecting mathematical expressions, sentences, and sequences that share common structural patterns, often focusing instead on incidental similarities or differences (Ericksen, 1991). Evidence suggests that these difficulties can be ameliorated by introducing early algebra ideas in the elementary grades (Carpenter, Franke, & Levi, 2003).

Research on the integration of new technologies into mathematics instruction continues to advance and is summarized in a National Research Council report (Kilpatrick, Swafford, & Findell, 2001). Kieran and Sfard (1999) discuss how students develop stronger conceptual knowledge of equations through graphing, and there is evidence that students can use technological tools to improve problem-solving even when their algebraic skills are limited (Huntley, Rasmussen, Villarubi, Sangtong, & Fey, 2000). Survey data suggest that middle- and high-school students think that mathematics has practical, everyday uses and that mathematics is more important for society than for them personally (Brown et al., 1988).

Typical student beliefs about mathematical inquiry that limit students' mathematical behavior (Schoenfeld, 1985, 2006) include the following: There is only one correct way to solve any mathematics problem and only one correct answer; mathematics is done by individuals in isolation; mathematical problems can be solved quickly or not at all and their solutions do not have to make sense; and formal proof is irrelevant to processes of discovery and invention (Schoenfeld, 1985, 1989a, 1989b). Research is needed to assess when and how students can understand that mathematical inquiry is a cycle in which ideas are represented abstractly, abstractions are manipulated, and results are tested against the original ideas. We must also learn at what age students can begin to represent something by a symbol or expression, and what standards students use to judge when solutions to mathematical problems are useful or adequate.

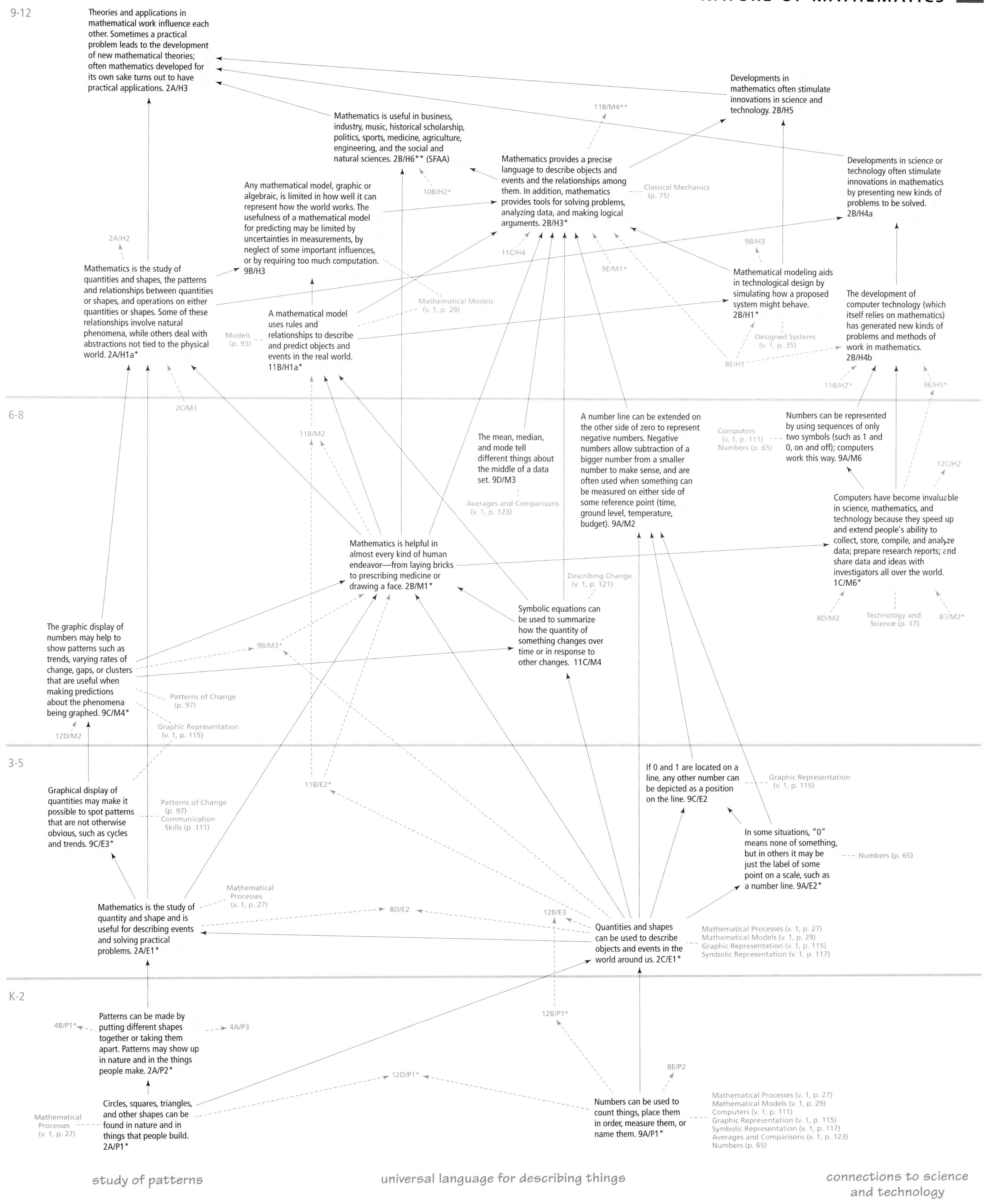
9-12
Theories and applications in mathematical work influence each other. Sometimes a practical problem leads to the development of new mathematical theories; often mathematics developed for its own sake turns out to have practical applications. 2A/H3
Developments in mathematics often stimulate innovations in science and technology. 2B/H5
11B/M4**
Mathematics is useful in business, industry, music, historical scholarship, politics, sports, medicine, agriculture, engineering, and the social and natural sciences. 2B/H6** (SFAA)
Mathematics provides a precise language to describe objects and events and the relationships among them. In addition, mathematics provides tools for solving problems, analyzing data, and making logical arguments. 2B/H3*
Classical Mechanics (p. 75)
Developments in science or technology often stimulate innovations in mathematics by presenting new kinds of problems to be solved. 2B/H4a
Any mathematical model, graphic or algebraic, is limited in how well it can represent how the world works. The usefulness of a mathematical model for predicting may be limited by uncertainties in measurements, by neglect of some important influences, or by requiring too much computation. 9B/H3
10B/H2*
2A/H2
11C/H4
9E/M1*
9B/H3
Mathematics is the study of quantities and shapes, the patterns and relationships between quantities or shapes, and operations on either quantities or shapes. Some of these relationships involve natural phenomena, while others deal with abstractions not tied to the physical world. 2A/H1a*
Mathematical Models (v. 1, p. 29)
Mathematical modeling aids in technological design by simulating how a proposed system might behave. 2B/H1*
The development of computer technology (which itself relies on mathematics) has generated new kinds of problems and methods of work in mathematics. 2B/H4b
Models (p. 93)
A mathematical model uses rules and relationships to describe and predict objects and events in the real world. 11B/H1a*
Designed Systems (v. 1, p. 35)
8E/H1
11B/H2*
9E/H5*
2C/M1
6-8
11B/M2
The mean, median, and mode tell different things about the middle of a data set. 9D/M3
A number line can be extended on the other side of zero to represent negative numbers. Negative numbers allow subtraction of a bigger number from a smaller number to make sense, and are often used when something can be measured on either side of some reference point (time, ground level, temperature, budget). 9A/M2
Computers (v. 1, p. 111)
Numbers (p. 65)
Numbers can be represented by using sequences of only two symbols (such as 1 and 0, on and off); computers work this way. 9A/M6
12C/H2
Averages and Comparisons (v. 1, p. 123)
Computers have become invaluable in science, mathematics, and technology because they speed up and extend people's ability to collect, store, compile, and analyze data; prepare research reports; and share data and ideas with investigators all over the world. 1C/M6*
Mathematics is helpful in almost every kind of human endeavor—from laying bricks to prescribing medicine or drawing a face. 2B/M1*
Describing Change (v. 1, p. 121)
8D/M2
Technology and Science (p. 17)
8E/M2*
The graphic display of numbers may help to show patterns such as trends, varying rates of change, gaps, or clusters that are useful when making predictions about the phenomena being graphed. 9C/M4*
9B/M3*
Symbolic equations can be used to summarize how the quantity of something changes over time or in response to other changes. 11C/M4
Patterns of Change (p. 97)
Graphic Representation (v. 1, p. 115)
12D/M2
3-5
If 0 and 1 are located on a line, any other number can be depicted as a position on the line. 9C/E2
Graphic Representation (v. 1, p. 115)
Graphical display of quantities may make it possible to spot patterns that are not otherwise obvious, such as cycles and trends. 9C/E3*
11B/E2*
Patterns of Change (p. 97)
Communication Skills (p. 111)
In some situations, "0" means none of something, but in others it may be just the label of some point on a scale, such as a number line. 9A/E2*
Numbers (p. 65)
Mathematical Processes (v. 1, p. 27)
Mathematics is the study of quantity and shape and is useful for describing events and solving practical problems. 2A/E1*
8D/E2
12B/E3
Quantities and shapes can be used to describe objects and events in the world around us. 2C/E1*
Mathematical Processes (v. 1, p. 27)
Mathematical Models (v. 1, p. 29)
Graphic Representation (v. 1, p. 115)
Symbolic Representation (v. 1, p. 117)
K-2
12B/P1*
4B/P1*
Patterns can be made by putting different shapes together or taking them apart. Patterns may show up in nature and in the things people make. 2A/P2*
4A/P3
8E/P2
12D/P1*
Mathematical Processes (v. 1, p. 27)
Circles, squares, triangles, and other shapes can be found in nature and in things that people build. 2A/P1*
Numbers can be used to count things, place them in order, measure them, or name them. 9A/P1*
Mathematical Processes (v. 1, p. 27)
Mathematical Models (v. 1, p. 29)
Computers (v. 1, p. 111)
Graphic Representation (v. 1, p. 115)
Symbolic Representation (v. 1, p. 117)
Averages and Comparisons (v. 1, p. 123)
Numbers (p. 65)
study of patterns
universal language for describing things
connections to science and technology

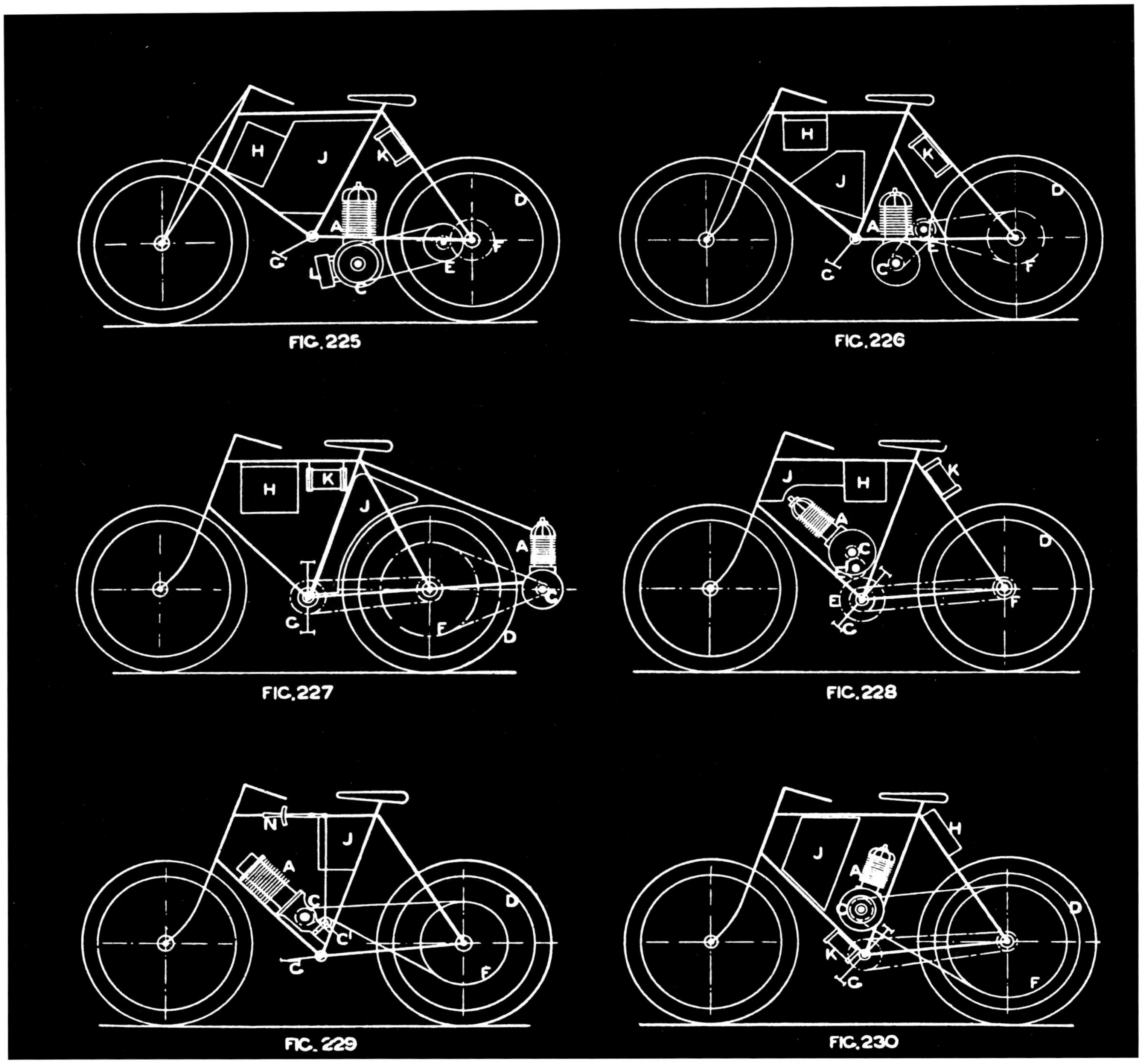


W. W. Beaumont *Alternative designs for motorized bicycles,* 1903

*As long as there have been people, there has been technology. Indeed, the techniques of shaping tools are taken as the chief evidence of the beginning of human culture. On the whole, technology has been a powerful force in the development of civilization, all the more so as its link with science has been forged. Technology—like language, ritual, values, commerce, and the arts—is an intrinsic part of a cultural system and it both shapes and reflects the system's values. In today's world, technology is a complex social enterprise that includes not only research, design, and crafts but also finance, manufacturing, management, labor, marketing, and maintenance.*

*In the broadest sense, technology extends our abilities to change the world: to cut, shape, or put together materials; to move things from one place to another; to reach farther with our hands, voices, and senses. We use technology to try to change the world to suit us better. The changes may relate to survival needs such as food, shelter, or defense, or they may relate to human aspirations such as knowledge, art, or control. But the results of changing the world are often complicated and unpredictable. They can include unexpected benefits, unexpected costs, and unexpected risks—any of which may fall on different social groups at different times. Anticipating the effects of technology is therefore as important as advancing its capabilities.*

# CHAPTER 3

# THE NATURE OF TECHNOLOGY



This chapter explores technology as a process, as a way of thinking, and as an enterprise that interacts with many other aspects of our world. Technology works in everyday life to shape the character of civilization. When people know how engineers go about their work and appreciate that all design operates within constraints, they are more likely to react thoughtfully to issues involving design tradeoffs and risks.

Beyond the maps presented in this chapter, a number of topics related to technology are treated in maps found in other chapters of *Atlas*. For example, Chapter 8: THE DESIGNED WORLD concerns itself with basic principles relevant to some of the key technologies of today's world; Chapter 10: HISTORICAL PERSPECTIVES suggests a sequence of ideas to help students understand the Industrial Revolution; Chapter 11: COMMON THEMES maps out powerful engineering concepts such as systems and models that are useful as analytical tools across disciplines; and Chapter 12: HABITS OF MIND identifies the skills needed for participating in a technological world.

The list on the left of this page shows the maps for this chapter and indicates the volume of *Atlas of Science Literacy* in which each appears. Those in green typeface are in *Volume 2* and those in gray are in *Volume 1*.

# TECHNOLOGY AND SCIENCE (3A)

Technology has been a powerful force in the development of civilization, all the more so as its link with science has been forged. Because of students' greater exposure to technology, it is especially important for them to understand its connection to science.

Like the scientific disciplines, the boundaries between science and technology are often blurred, so a clear distinction between them is not the point. Rather, students should understand how technology draws on science and contributes to it.

The map is organized around three strands—*effect of technology on science, effect of science on technology,* and *effect on society*. The progression of understanding begins in the elementary grades with a focus on using tools and appreciating their value, not just in science class but in everyday life. Later grades emphasize contributions of various technologies to society and kinds of interactions among science and technology.

Because technology and science are so closely related to mathematics, users of this map will find it helpful to refer to the **NATURE OF MATHEMATICS** map. Specific technologies described in maps from Chapter 8: THE DESIGNED WORLD can be used to illustrate the ideas on this map. Additionally, middle-school benchmarks that are part of the **GALAXIES AND THE UNIVERSE** map in *Atlas 1* can provide contexts for learning about how technology affects science. At the high-school level, maps in Chapter 10: HISTORICAL PERSPECTIVES describe discoveries and developments involving technology and science that provide interesting contexts for all three strands on this map.

## NOTES

The strand *effect of technology on science* is designed to show how students might advance beyond the common notion of "technology as handmaiden to science" and begin to understand the synergistic nature of the relationship between the two disciplines. Although the map alludes to the telescope—and the discoveries it makes possible—as an example of the impact of technology on science, a variety of other examples such as imaging technologies and computer modeling may be used to help students appreciate that technology impacts *all* of science. Several historical episodes, such as **THE INDUSTRIAL REVOLUTION**, are mapped in Chapter 10 and can be used to illustrate that technological advances contribute directly to scientific progress. The new 9-12 benchmark in the *effect of science on technology* strand is from *Science for All Americans*. High school benchmarks on the **SPLITTING THE ATOM** and **DISCOVERING GERMS** maps can also contribute to an understanding of the contribution of scientific knowledge to technological discovery.

The 9-12 benchmark in the *effect on society* strand indicates that students are to understand the different roles of science and technology and their somewhat different effects on society. These distinctions are developed in more depth in the **SCIENCE AND SOCIETY** map in this volume and in the **INTERACTION OF TECHNOLOGY AND SOCIETY** map in *Atlas 1*.

### RESEARCH IN *BENCHMARKS*

Even in middle school, students typically do not distinguish between an engineering model of experimentation, where the goal is to produce a desirable outcome, and the scientific model of experimentation, where the goal is to understand the relation between causes and effects (Carey et al., 1989; Schauble et al., 1991). Some research suggests that students can understand and use the engineering model before they can the scientific model—that is, that students inevitably will think about producing desirable outcomes before they are able to do the more analytic form of thinking involved in scientific inquiry (Schauble et al., 1991).

High-school students do not distinguish between the roles of science and technology unless explicitly asked to do so (Fleming, 1987). This is evidenced, for example, by students' view that science serves the public interest. More generally, some students believe science affects society in more positive ways than does technology. That is partly because students associate science with medical research but associate technology with pollution or weapons. Students appear to understand the impact of science on technology, but they do not always appreciate the impact of technology on science (Fleming, 1987).

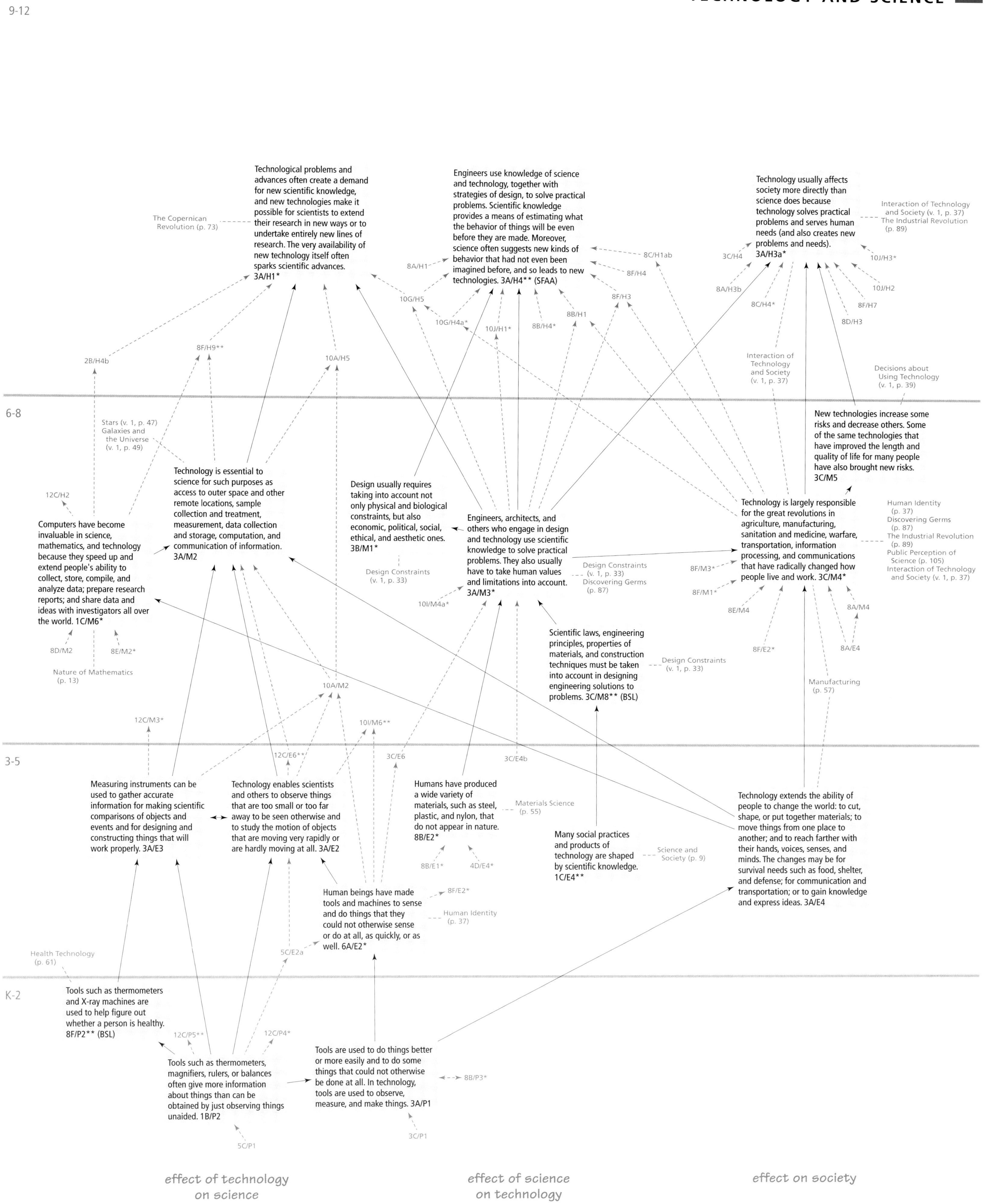
9-12
6-8
3-5
K-2
Technological problems and advances often create a demand for new scientific knowledge, and new technologies make it possible for scientists to extend their research in new ways or to undertake entirely new lines of research. The very availability of new technology itself often sparks scientific advances. 3A/H1*
The Copernican Revolution (p. 73)
Engineers use knowledge of science and technology, together with strategies of design, to solve practical problems. Scientific knowledge provides a means of estimating what the behavior of things will be even before they are made. Moreover, science often suggests new kinds of behavior that had not even been imagined before, and so leads to new technologies. 3A/H4** (SFAA)
Technology usually affects society more directly than science does because technology solves practical problems and serves human needs (and also creates new problems and needs). 3A/H3a*
Interaction of Technology and Society (v. 1, p. 37)
The Industrial Revolution (p. 89)
8C/H1ab
8A/H1
8F/H4
8F/H3
8B/H1
10G/H5
10G/H4a*
10J/H1*
8B/H4*
3C/H4
10J/H3*
10J/H2
8A/H3b
8C/H4*
8F/H7
8D/H3
2B/H4b
8F/H9**
10A/H5
Interaction of Technology and Society (v. 1, p. 37)
Decisions about Using Technology (v. 1, p. 39)
Stars (v. 1, p. 47)
Galaxies and the Universe (v. 1, p. 49)
New technologies increase some risks and decrease others. Some of the same technologies that have improved the length and quality of life for many people have also brought new risks. 3C/M5
12C/H2
Technology is essential to science for such purposes as access to outer space and other remote locations, sample collection and treatment, measurement, data collection and storage, computation, and communication of information. 3A/M2
Design usually requires taking into account not only physical and biological constraints, but also economic, political, social, ethical, and aesthetic ones. 3B/M1*
Engineers, architects, and others who engage in design and technology use scientific knowledge to solve practical problems. They also usually have to take human values and limitations into account. 3A/M3*
Technology is largely responsible for the great revolutions in agriculture, manufacturing, sanitation and medicine, warfare, transportation, information processing, and communications that have radically changed how people live and work. 3C/M4*
Human Identity (p. 37)
Discovering Germs (p. 87)
The Industrial Revolution (p. 89)
Public Perception of Science (p. 105)
Interaction of Technology and Society (v. 1, p. 37)
Computers have become invaluable in science, mathematics, and technology because they speed up and extend people's ability to collect, store, compile, and analyze data; prepare research reports; and share data and ideas with investigators all over the world. 1C/M6*
Design Constraints (v. 1, p. 33)
Design Constraints (v. 1, p. 33)
Discovering Germs (p. 87)
8F/M3*
8F/M1*
10I/M4a*
8E/M4
8A/M4
8D/M2
8E/M2*
Nature of Mathematics (p. 13)
Scientific laws, engineering principles, properties of materials, and construction techniques must be taken into account in designing engineering solutions to problems. 3C/M8** (BSL)
Design Constraints (v. 1, p. 33)
8F/E2*
8A/E4
Manufacturing (p. 57)
10A/M2
12C/M3*
10I/M6**
12C/E6**
3C/E6
3C/E4b
Measuring instruments can be used to gather accurate information for making scientific comparisons of objects and events and for designing and constructing things that will work properly. 3A/E3
Technology enables scientists and others to observe things that are too small or too far away to be seen otherwise and to study the motion of objects that are moving very rapidly or are hardly moving at all. 3A/E2
Humans have produced a wide variety of materials, such as steel, plastic, and nylon, that do not appear in nature. 8B/E2*
Materials Science (p. 55)
Technology extends the ability of people to change the world: to cut, shape, or put together materials; to move things from one place to another; and to reach farther with their hands, voices, senses, and minds. The changes may be for survival needs such as food, shelter, and defense; for communication and transportation; or to gain knowledge and express ideas. 3A/E4
Many social practices and products of technology are shaped by scientific knowledge. 1C/E4**
Science and Society (p. 9)
8B/E1*
4D/E4*
Human beings have made tools and machines to sense and do things that they could not otherwise sense or do at all, or as quickly, or as well. 6A/E2*
8F/E2*
Human Identity (p. 37)
Health Technology (p. 61)
5C/E2a
Tools such as thermometers and X-ray machines are used to help figure out whether a person is healthy. 8F/P2** (BSL)
12C/P5**
12C/P4*
Tools such as thermometers, magnifiers, rulers, or balances often give more information about things than can be obtained by just observing things unaided. 1B/P2
Tools are used to do things better or more easily and to do some things that could not otherwise be done at all. In technology, tools are used to observe, measure, and make things. 3A/P1
8B/P3*
5C/P1
3C/P1
effect of technology on science
effect of science on technology
effect on society

Thomas Hart Benton *Trail Riders*, 1964-65

*Humans have never lost interest in trying to find out how the universe is put together, how it works, and where they fit in the cosmic scheme of things. The development of our understanding of the architecture of the universe is surely not complete, but we have made great progress. Given a universe that is made up of distances too vast to reach and of particles too small to see and too numerous to count, it is a tribute to human intelligence that we have made as much progress as we have in accounting for how things fit together. All humans should participate in the pleasure of coming to know their universe better.*

*This chapter consists of recommendations for basic knowledge about the overall structure of the universe and the physical principles on which it seems to run, with emphasis on the earth and the solar system. The chapter focuses on two principal subjects: the structure of the universe and the major processes that have shaped the planet earth, and the concepts with which science describes the physical world in general—organized for convenience under the headings of matter, energy, motion, and forces.*

CHAPTER 4

# THE PHYSICAL SETTING



The scientific effort to understand the universe and how it operates is part of an enduring human imperative. One of the grand success stories of science is the unification of the physical universe. It turns out that all natural objects, events, and processes are connected to each other in such a way that only a relatively few concepts are needed to make sense of them. This chapter describes basic principles of astronomy, earth science, chemistry, and physics that provide a basis for making sense of phenomena in the physical setting and serve as a foundation for learning more. In addition, maps in Chapter 10: HISTORICAL PERSPECTIVES provide accounts of significant discoveries and changes that revolutionized our understanding of the physical world and how it operates.

The list on the left of this page shows the maps for this chapter and indicates the volume of *Atlas of Science Literacy* in which each appears. Those in green typeface are in *Volume 2* and those in gray are in *Volume 1*.

# WEATHER AND CLIMATE (4B)

The earth has a variety of climatic patterns, which consist of different conditions of temperature, precipitation, humidity, wind, air pressure, and other atmospheric phenomena. These result from a variety of factors. Climate and changes in climate have influenced in the past and will continue to influence what kinds of life forms are able to exist. Understanding the basic principles that contribute to maintaining and causing changes in weather and climate increases our ability to forecast and moderate the effects of weather and to make informed decisions about human activities that may contribute to climate change.

The map is organized around four strands—*temperature and winds, water cycle, atmosphere,* and *climate change*. The progression of understanding begins in the elementary grades with observations about heat transfer, changes in water from one state to another, and changes in weather over the course of a day and over the course of seasons. By middle school, the focus is on the water cycle, patterns of change in temperature, and the notion of climate change. In high school, seasons and winds and the water cycle are related to gravity and the earth's rotation, and climate change is related to natural causes and human activities.

Benchmarks in this map about temperature and winds draw on ideas about heat transfer and transformation in the **ENERGY TRANSFORMATIONS** map. Benchmarks in the *climate change* strand are also related to the **SCIENCE AND SOCIETY** map. The widespread use of climate models to improve our understanding of the earth's climate system and climate change suggests a connection to benchmarks in the **MODELS** map as well.

## NOTES

The left-hand side of the *temperature and winds* strand presents a progression of understanding of seasons. The explanation of the seasons in terms of the tilt of the earth requires students to engage in fairly complex spatial reasoning. For this reason, although the idea is introduced at the 6-8 grade level in *Benchmarks*, the map places it (4B/H3) at the 9-12 level.

Benchmarks related to the heating of materials and the transfer of thermal energy lay the conceptual groundwork for understanding solar heating, global circulation, seasonal weather patterns and climate, and the effect of greenhouse gases. To understand how thermal energy moves in both oceanic and atmospheric systems, students need to know that convective currents are an essential mechanism that aids in that movement. In middle school, understanding of convection currents is linked to experiences with relevant phenomena. Understanding convection in terms of gravity, buoyant forces, and pressure is not expected until high school. It is not necessary for students to have a molecular comprehension of thermal energy to be able to understand atmospheric and oceanic circulation patterns and their role in climate.

Several lines of conceptual development converge in the new 9-12 benchmark that begins "Climatic conditions result from...." These include an understanding of temperature patterns over the earth, atmospheric and oceanic circulation patterns, and the water cycle. A double-headed arrow between this benchmark and another new benchmark (4B/H6) on climate change indicates that they are closely related but that neither is conceptually dependent on the other.

## RESEARCH IN *BENCHMARKS*

Students of all ages (including college students and adults) have difficulty understanding what causes the seasons. Students may not be able to understand explanations of the seasons before they reasonably understand the relative size, motion, and distance of the sun and the earth (Sadler, 1987; Vosniadou, 1991). Many students before and after instruction in earth science think that winter is colder than summer because the earth is farther from the sun in winter (Atwood & Atwood, 1996; Dove, 1998; Philips, 1991; Sadler, 1998). This idea is often related to the belief that the earth orbits the sun in an elongated elliptical path (Galili & Lavrik, 1998; Sadler, 1998). Other students, especially after instruction, think that the distance between the northern hemisphere and the sun changes because the earth leans toward the sun in the summer and away from the sun in winter (Galili & Lavrik, 1998; Sadler, 1998). Students' ideas about how light travels and about the earth-sun relationship, including the shape of the earth's orbit, the period of the earth's revolution around the sun, and the period of the earth's rotation around its axis, may interfere with students' understanding of the seasons (Galili & Lavrik, 1998; Salierno, Edelson, & Sherin, 2005). For example, some students believe that the side of the earth not facing the sun experiences winter, indicating a confusion between the daily rotation of the earth and its yearly revolution around the sun (Salierno, Edelson, & Sherin, 2005).

Although upper elementary students may identify air as existing even in static situations and recognize that it takes space, recognizing that air has weight may be challenging even for high-school students (Sere, 1985; Driver et al., 1994a; Krnel, Watson, & Glazar, 1998). Students of all ages (including college students) may believe that air exerts force or pressure only when it is moving and only downwards (Driver et al., 1994a; Sere, 1985; Henriques, 2002; Nelson, Aron, & Francek, 1992). Only a few middle-school students use the idea of pressure differences between regions of the atmosphere to account for wind; instead, they may account for winds in terms of visible moving objects or the movement of the earth (Driver et al., 1994a).

Before students understand that water is converted to an invisible form, they may initially believe that when water evaporates it ceases to exist, or that it changes location but remains a liquid, or that it is transformed into some other perceptible form (fog, steam, droplets, etc.) (Bar, 1989; Russell, Harlen, & Watt, 1989; Russell & Watt, 1990; Krnel, Watson, & Glazar, 1998). With special instruction, some students in 5th grade may be able to identify the air as the final location of evaporating water (Russell & Watt, 1990), but they must first accept air as a permanent substance (Bar, 1989). For many students, difficulty understanding the existence of water vapor in the atmosphere persists in middle school years (Lee et al., 1993; Johnson, 1998). Students can understand rainfall in terms of gravity once they attribute weight to little drops of water (typically in upper elementary grades), but the mechanism through which condensation occurs may not be understood until high school (Bar, 1989).

Students of all ages may confuse the ozone layer with the greenhouse effect, and may have a tendency to imagine that all environmentally friendly actions help to solve all environmental problems (for example, that the use of unleaded petrol reduces the risk of global warming) (Andersson & Wallin, 2000; Koulaidis & Christidou, 1998; Meadows & Wiesenmayer, 1999; Rye, Rubba, & Wiesenmayer, 1997). Students have difficulty linking relevant elements of knowledge when explaining the greenhouse effect and may confuse the natural greenhouse effect with the enhancement of that effect (Andersson & Wallin, 2000).

See **ENERGY RESOURCES** and **ENERGY TRANSFORMATIONS** for additional research.

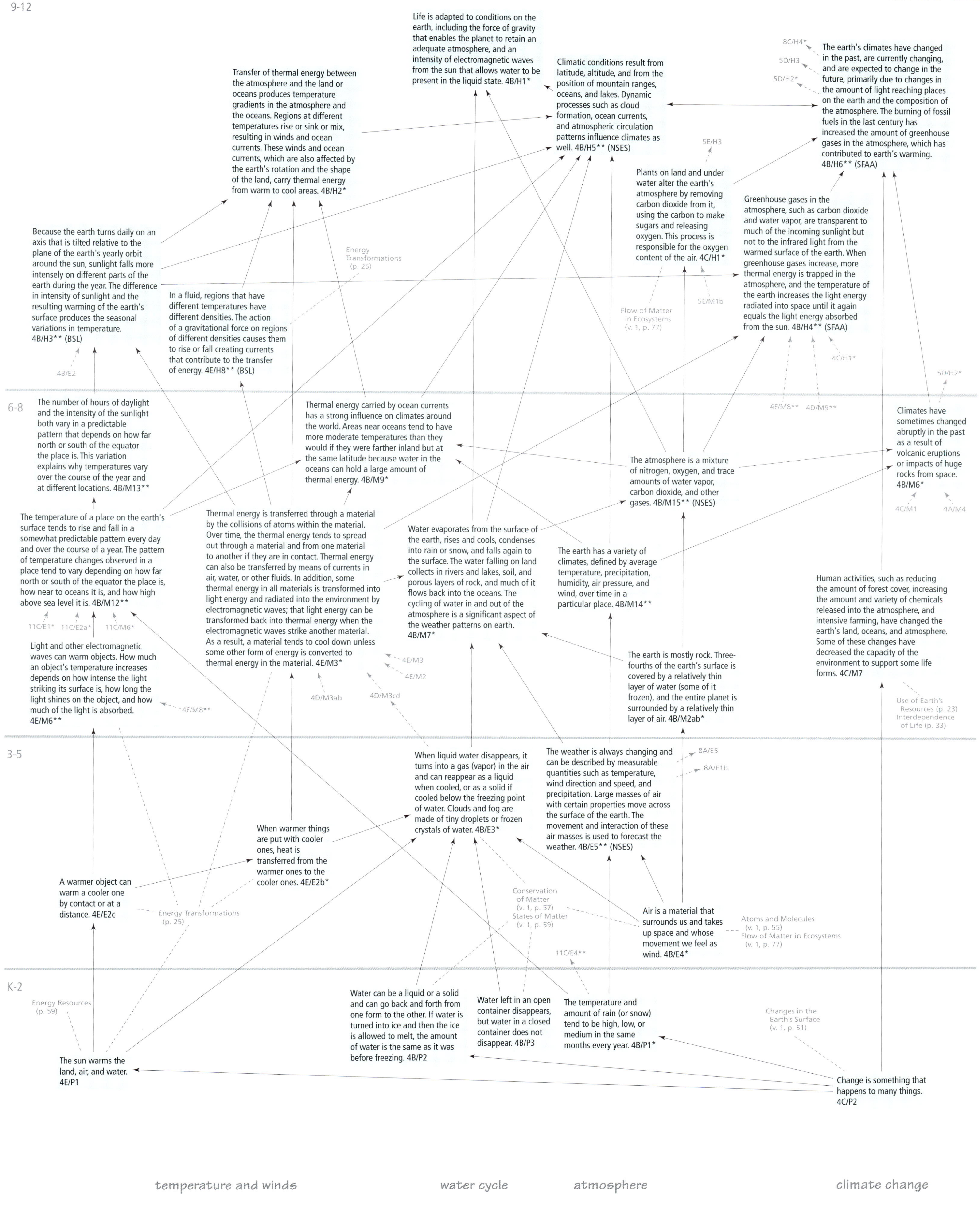
9-12
Life is adapted to conditions on the earth, including the force of gravity that enables the planet to retain an adequate atmosphere, and an intensity of electromagnetic waves from the sun that allows water to be present in the liquid state. 4B/H1*
Transfer of thermal energy between the atmosphere and the land or oceans produces temperature gradients in the atmosphere and the oceans. Regions at different temperatures rise or sink or mix, resulting in winds and ocean currents. These winds and ocean currents, which are also affected by the earth's rotation and the shape of the land, carry thermal energy from warm to cool areas. 4B/H2*
Climatic conditions result from latitude, altitude, and from the position of mountain ranges, oceans, and lakes. Dynamic processes such as cloud formation, ocean currents, and atmospheric circulation patterns influence climates as well. 4B/H5** (NSES)
8C/H4*
5D/H3
5D/H2*
The earth's climates have changed in the past, are currently changing, and are expected to change in the future, primarily due to changes in the amount of light reaching places on the earth and the composition of the atmosphere. The burning of fossil fuels in the last century has increased the amount of greenhouse gases in the atmosphere, which has contributed to earth's warming. 4B/H6** (SFAA)
5E/H3
Plants on land and under water alter the earth's atmosphere by removing carbon dioxide from it, using the carbon to make sugars and releasing oxygen. This process is responsible for the oxygen content of the air. 4C/H1*
Because the earth turns daily on an axis that is tilted relative to the plane of the earth's yearly orbit around the sun, sunlight falls more intensely on different parts of the earth during the year. The difference in intensity of sunlight and the resulting warming of the earth's surface produces the seasonal variations in temperature. 4B/H3** (BSL)
Energy Transformations (p. 25)
In a fluid, regions that have different temperatures have different densities. The action of a gravitational force on regions of different densities causes them to rise or fall creating currents that contribute to the transfer of energy. 4E/H8** (BSL)
Greenhouse gases in the atmosphere, such as carbon dioxide and water vapor, are transparent to much of the incoming sunlight but not to the infrared light from the warmed surface of the earth. When greenhouse gases increase, more thermal energy is trapped in the atmosphere, and the temperature of the earth increases the light energy radiated into space until it again equals the light energy absorbed from the sun. 4B/H4** (SFAA)
Flow of Matter in Ecosystems (v. 1, p. 77)
5E/M1b
4C/H1*
4B/E2
5D/H2*
6-8
The number of hours of daylight and the intensity of the sunlight both vary in a predictable pattern that depends on how far north or south of the equator the place is. This variation explains why temperatures vary over the course of the year and at different locations. 4B/M13**
Thermal energy carried by ocean currents has a strong influence on climates around the world. Areas near oceans tend to have more moderate temperatures than they would if they were farther inland but at the same latitude because water in the oceans can hold a large amount of thermal energy. 4B/M9*
4F/M8**
4D/M9**
Climates have sometimes changed abruptly in the past as a result of volcanic eruptions or impacts of huge rocks from space. 4B/M6*
The atmosphere is a mixture of nitrogen, oxygen, and trace amounts of water vapor, carbon dioxide, and other gases. 4B/M15** (NSES)
4C/M1
4A/M4
The temperature of a place on the earth's surface tends to rise and fall in a somewhat predictable pattern every day and over the course of a year. The pattern of temperature changes observed in a place tend to vary depending on how far north or south of the equator the place is, how near to oceans it is, and how high above sea level it is. 4B/M12**
Thermal energy is transferred through a material by the collisions of atoms within the material. Over time, the thermal energy tends to spread out through a material and from one material to another if they are in contact. Thermal energy can also be transferred by means of currents in air, water, or other fluids. In addition, some thermal energy in all materials is transformed into light energy and radiated into the environment by electromagnetic waves; that light energy can be transformed back into thermal energy when the electromagnetic waves strike another material. As a result, a material tends to cool down unless some other form of energy is converted to thermal energy in the material. 4E/M3*
Water evaporates from the surface of the earth, rises and cools, condenses into rain or snow, and falls again to the surface. The water falling on land collects in rivers and lakes, soil, and porous layers of rock, and much of it flows back into the oceans. The cycling of water in and out of the atmosphere is a significant aspect of the weather patterns on earth. 4B/M7*
The earth has a variety of climates, defined by average temperature, precipitation, humidity, air pressure, and wind, over time in a particular place. 4B/M14**
Human activities, such as reducing the amount of forest cover, increasing the amount and variety of chemicals released into the atmosphere, and intensive farming, have changed the earth's land, oceans, and atmosphere. Some of these changes have decreased the capacity of the environment to support some life forms. 4C/M7
11C/E1*
11C/E2a*
11C/M6*
Light and other electromagnetic waves can warm objects. How much an object's temperature increases depends on how intense the light striking its surface is, how long the light shines on the object, and how much of the light is absorbed. 4E/M6**
4F/M8**
4E/M3
4E/M2
4D/M3ab
4D/M3cd
The earth is mostly rock. Three-fourths of the earth's surface is covered by a relatively thin layer of water (some of it frozen), and the entire planet is surrounded by a relatively thin layer of air. 4B/M2ab*
Use of Earth's Resources (p. 23)
Interdependence of Life (p. 33)
3-5
When liquid water disappears, it turns into a gas (vapor) in the air and can reappear as a liquid when cooled, or as a solid if cooled below the freezing point of water. Clouds and fog are made of tiny droplets or frozen crystals of water. 4B/E3*
The weather is always changing and can be described by measurable quantities such as temperature, wind direction and speed, and precipitation. Large masses of air with certain properties move across the surface of the earth. The movement and interaction of these air masses is used to forecast the weather. 4B/E5** (NSES)
8A/E5
8A/E1b
When warmer things are put with cooler ones, heat is transferred from the warmer ones to the cooler ones. 4E/E2b*
A warmer object can warm a cooler one by contact or at a distance. 4E/E2c
Energy Transformations (p. 25)
Conservation of Matter (v. 1, p. 57)
States of Matter (v. 1, p. 59)
Air is a material that surrounds us and takes up space and whose movement we feel as wind. 4B/E4*
Atoms and Molecules (v. 1, p. 55)
Flow of Matter in Ecosystems (v. 1, p. 77)
11C/E4**
K-2
Energy Resources (p. 59)
Water can be a liquid or a solid and can go back and forth from one form to the other. If water is turned into ice and then the ice is allowed to melt, the amount of water is the same as it was before freezing. 4B/P2
Water left in an open container disappears, but water in a closed container does not disappear. 4B/P3
The temperature and amount of rain (or snow) tend to be high, low, or medium in the same months every year. 4B/P1*
Changes in the Earth's Surface (v. 1, p. 51)
The sun warms the land, air, and water. 4E/P1
Change is something that happens to many things. 4C/P2
temperature and winds
water cycle
atmosphere
climate change

## THE PHYSICAL SETTING

# USE OF EARTH'S RESOURCES (4B)

Of all the diverse planets and moons in our solar system, only the earth appears to be capable of supporting life as we know it. Water and air are essential resources; altering the quality of the water or the composition of the earth's atmosphere can have serious consequences for living systems. Informed citizens need to understand the implications of decisions about the use of Earth's resources.

The map is organized around four strands—*use of energy resources, needs of organisms for Earth's resources, human impact on the environment,* and *use of material resources*. In the elementary grades, the focus is on the survival needs of organisms and people's use of fuels. In middle school, the focus is on the use and trade-offs of different energy and material resources and on the dependence and impact of organisms (particularly humans) on their environment. In high school, the focus is on factors affecting the equilibrium of the synthesis and depletion of resources and the recycling of waste products resulting from resource use.

This map is closely related to both the **ENERGY RESOURCES** and **MATERIALS SCIENCE** maps.

**NOTES**

The *use of energy resources* strand focuses on the increased use and depletion of energy resources. Benchmarks in this strand also appear in the **ENERGY RESOURCES** map. Most of the benchmarks in the *use of material resources* strand also appear in the **MATERIALS SCIENCE** map, which addresses in more depth the problem of waste disposal and recycling.

The *use of energy resources* and *use of material resources* strands include many similar key ideas: Some resources are renewable, some are not, and some are renewable at great cost; some resources take a very long time to accumulate; the growth of technology has led to increased use of resources; as resources are depleted, they may be more difficult and costly to obtain; and use of resources is associated with environmental risks.

The *needs of organisms for Earth's resources* strand addresses how organisms depend on their environment for resources, and shares several benchmarks with the **INTERDEPENDENCE OF LIFE** map. The *human impact on the environment* strand considers how humans affect the environment. The two strands converge on the 9-12 benchmark "Human beings are part of the earth's ecosystems...."

Students' understanding of the use of Earth's resources should be integrated with their growing knowledge of the processes that shape the earth (see the *Atlas 1* map **CHANGES IN THE EARTH'S SURFACE**). For example, knowing about the slow nature of the rock cycle would enable students to understand that mineral resources cannot be renewed for human use within a time span that is determined by the rate at which humans deplete them.

**RESEARCH IN *BENCHMARKS***

No relevant research available in *Benchmarks*.

# USE OF EARTH'S RESOURCES

use of energy resources | needs of organisms for Earth's resources | human impact on the environment | use of material resources

## K-2

People burn fuels such as wood, oil, coal, or natural gas, or use electricity, to cook their food and warm their houses. 8C/P2

Energy Resources (p. 59)

Plants and animals both need to take in water, and animals need to take in food. In addition, plants need light. 5E/P1

Flow of Matter in Ecosystems (v. 1, p. 77)
Flow of Energy in Ecosystems (v. 1, p. 79)

People need water, food, air, waste removal, and a particular range of temperatures in their environment, just as other animals do. 6A/P2

Cell Functions (v. 1, p. 73)
Human Identity (p. 37)
Basic Functions (p. 41)

## 3-5

Some people try to reduce the amount of fuels they use in order to conserve resources, reduce pollution, or to save money. 8C/E4*

For any particular environment, some kinds of plants and animals thrive, some do not live as well, and some do not survive at all. 5D/E1*

Interdependence of Life (p. 33)

Naturally occurring materials such as wood, clay, cotton, and animal skins may be processed to change their properties. 8B/E1*

8C/E2*
Natural Selection (v. 1, p. 83)
Materials Science (p. 55)

## 6-8

Industry, transportation, urban development, agriculture, and most other human activities are closely tied to the amount and kind of energy available. People in different parts of the world have different amounts and kinds of energy resources to use and use them for different purposes. 8C/M6*

Some resources are not renewable or renew very slowly. Fuels already accumulated in the earth, for instance, will become more difficult to obtain as the most readily available resources run out. How long the resources will last, however, is difficult to predict. The ultimate limit may be the prohibitive cost of obtaining them. 8C/M10** (SFAA)

Different ways of obtaining, transforming, and distributing energy have different environmental consequences. 8C/M2

Energy from the sun (and the wind and water energy derived from it) is available indefinitely. Because the transfer of energy from these resources is weak and variable, systems are needed to collect and concentrate the energy. 8C/M5*

Energy Resources (p. 59)

Given adequate resources and an absence of disease or predators, populations of organisms in ecosystems increase at rapid rates. Finite resources and other factors limit their growth. 5D/M3** (NSES)

In all environments, organisms with similar needs may compete with one another for limited resources, including food, space, water, air, and shelter. 5D/M1a*

Interdependence of Life (p. 33)

The world contains a wide diversity of physical conditions, which creates a wide variety of environments: freshwater, marine, forest, desert, grassland, mountain, and others. In any particular environment, the growth and survival of organisms depend on the physical conditions. 5D/M1b*

The benefits of Earth's resources—such as fresh water, air, soil, and trees—can be reduced by deliberately or inadvertently polluting them. The atmosphere, the oceans, and the land have a limited capacity to absorb and recycle waste materials. In addition, some materials take a long time to degrade. Therefore, cleaning up polluted air, water, or soil can be difficult and costly. 4B/M11bc*

Human activities, such as reducing the amount of forest cover, increasing the amount and variety of chemicals released into the atmosphere, and intensive farming, have changed the earth's land, oceans, and atmosphere. Some of these changes have decreased the capacity of the environment to support some life forms. 4C/M7

Fresh water, limited in supply, is essential for some organisms and industrial processes. Water in rivers, lakes, and underground can be depleted or polluted, making it unavailable or unsuitable for life. 4B/M8*

Weather and Climate (p. 21)
Interdependence of Life (p. 33)

The wasteful or unnecessary use of natural resources can limit their availability for other purposes. Restoring depleted soil, forests, or fishing grounds can be difficult and costly. 4B/M11a*

Recycling materials and the development of substitutes for those materials can reduce the rate of depletion of resources but may also be costly. Some materials are not easily recycled. 4B/M10c*

Some material resources are very rare and some exist in great quantities. The ability to obtain and process resources depends on where they are located and the form they are in. As resources are depleted, they may become more difficult to obtain. 4B/M10ab*

Materials Science (p. 55)

7G/M1* 7G/M5* 8C/M11** 6E/M5 7G/M5* 7G/M1* 4B/M7* 7G/M5*

## 9-12

Decisions to slow the depletion of energy resources can be made at many levels, from personal to national, and they always involve trade-offs involving economic costs and social values. 8C/H5*

Energy Resources (p. 59)
8C/H2*

Industrialization brings an increased demand for and use of energy. Such usage contributes to having many more goods and services in the industrially developing nations but also leads to more rapid depletion of the earth's energy resources and to environmental risks associated with some energy resources. 8C/H4*

4B/H6** 8C/H3* 5E/H1*

Although the earth has a great capacity to absorb and recycle materials naturally, ecosystems have only a finite capacity to withstand change without experiencing major ecological alterations that may also have adverse effects on human activities. 4B/H9** (SFAA)

Human beings are part of the earth's ecosystems. Human activities can, deliberately or inadvertently, alter the equilibrium in ecosystems. 5D/H3

Ecosystems can be reasonably stable over hundreds or thousands of years. As any population grows, its size is limited by one or more environmental factors: availability of food, availability of nesting sites, or number of predators. 5D/H1*

Weather and Climate (p. 21)
4B/M2cd
Interdependence of Life (p. 33)

The human species has a major impact on other species in many ways: reducing the amount of the earth's surface available to those other species, interfering with their food sources, changing the temperature and chemical composition of their habitats, introducing foreign species into their ecosystems, and altering organisms directly through selective breeding and genetic engineering. 3C/H4

The earth has many natural resources of great importance to human life. Some are readily renewable, some are renewable only at great cost, and some are not renewable at all. 4B/H8** (SFAA)

Manufacturing (p. 57)
Materials Science (p. 55)

The development of new materials and the increased use of existing materials by a growing human population have led to the removal of resources from the environment much more rapidly than they can be replaced by natural processes. Disposal of waste materials has also become a problem. Solving these problems requires systematic efforts involving both social and technological innovations. 8B/H7** (SFAA)

# ENERGY TRANSFORMATIONS (4E)

Most of what goes on in the universe—such as the collapsing and exploding of stars, biological growth and decay, the operation of machines and computers—involves transforming one form of energy into another. Understanding basic ideas about energy transformation, including its conservation and dissipation, is needed for making sense of a wide variety of phenomena and for making informed decisions involving energy use.

The map is organized around three strands—*forms of energy, conservation,* and *dissipation*. The progression of understanding begins in elementary grades with observable patterns of phenomena involving heat transfer; moves in middle school to descriptions of energy forms, transformation, and transfer; and culminates in high school with energy conservation and dissipation as thermal energy. While there are really only two forms of energy—kinetic and potential—students typically encounter energy forms with more varied labels. Hence, the strand on forms of energy begins in middle school with a more diverse list of forms and reconciles them in high school into kinetic or potential.

Ideas about energy conservation on this map relate to the more general idea of conservation developed in the **CONSTANCY** map, and ideas about energy transformation contribute to understanding ideas about energy transformation in living systems on the **FLOW OF ENERGY IN ECOSYSTEMS** map in *Atlas 1*. Ideas about heat transfer contribute to understanding ideas about temperature and winds on the **WEATHER AND CLIMATE** map, and ideas about energy released in nuclear reactions contribute to understanding the history of ideas on the **RELATIVITY** and **SPLITTING THE ATOM** maps. Benchmarks in this map also contribute to and are supported by benchmarks in the **ENERGY RESOURCES** map.

## NOTES

In the *forms of energy* strand a new high-school benchmark "Many forms of energy..." synthesizes distinctions between kinetic energy (which concerns the motion of objects in a system) and potential energy (which concerns the relative position of interacting objects in a system).

The *conservation* strand begins with the precursor idea of accounting, keeping track of where things come from and where they go. The idea of energy conservation is introduced first qualitatively in benchmark 4E/M1. The second part of the benchmark was added to help address a common student misconception about energy loss. The high-school benchmark 4E/H1 expresses concretely the idea of energy conservation quantitatively: If we can keep track of how much energy of each kind increases and decreases, we find that whenever the energy in one place diminishes, the energy in other places increases by just the same amount. A new high-school benchmark "If no energy is transferred into or out of a system..." expresses the same idea more abstractly.

Describing processes in terms of energy transformation, transfer, and conservation in any meaningful way requires first defining the system under consideration. A new benchmark 11A/M5, shown as an off-map connection, expects students to know that a system is "defined by placing boundaries around collections of interrelated things," and since systems often interact with their environment, students should also know that "it is important to keep track of what enters or leaves the system."

## RESEARCH IN *BENCHMARKS*

Even after some years of physics instruction, students do not distinguish well between heat and temperature when they explain thermal phenomena (Kesidou & Duit, 1993; Tiberghien, 1983; Wiser, 1986, 1988). Their belief that temperature is the measure of heat is particularly resistant to change. Long-term teaching interventions are required for upper middle-school students to start differentiating between heat and temperature (Linn & Songer, 1991; Clark & Linn, 2003).

Few middle- and high-school students understand the molecular basis of heat conduction even after instruction (Wiser, 1986; Kesidou & Duit, 1993; Driver et al., 1994). For example, students attribute to particles properties such as "hotness" and "coldness" or believe that heat is produced by particles rubbing against each other (Driver et al., 1994; Kesidou, Duit, & Glynn, 1995).

During instruction, upper elementary-school students use ideas that give heat an active drive or intent to explain observations of convection currents. They also draw parallels between evaporation and the water cycle and convection, sometimes explicitly explaining the upwards motion of convection currents as evaporation (Jones, Carter, & Rua, 2000).

Students rarely think energy is measurable and quantifiable (Solomon, 1985; Watts, 1983a). Students' alternative conceptualizations of energy influence their interpretations of textbook representations of energy (Stylianidou, Ormerod, & Ogborn, 2002).

Middle- and high-school students tend to think that energy transformations involve only one form of energy at a time (Brook & Wells, 1988). Although they develop some skill in identifying different forms of energy, in most cases their descriptions of energy-change focus only on forms which have perceivable effects (Brook & Driver, 1986). The transformation of motion to heat seems to be difficult for students to accept, especially in cases with no temperature increase (Brook & Driver, 1986; Kesidou & Duit, 1993). Finally, it may not be clear to students that some forms of energy, such as light, sound, and chemical energy, can be used to make things happen (Carr & Kirkwood, 1988).

The idea of energy conservation seems counterintuitive to middle- and high-school students who hold on to the everyday use of the term energy, but teaching heat dissipation ideas at the same time as energy conservation ideas may help alleviate this difficulty (Solomon, 1983). Even after instruction, however, students do not seem to appreciate that energy conservation is a useful way to explain phenomena (Brook & Driver, 1984). A key difficulty students have in understanding conservation appears to derive from not considering the appropriate system and environment (Ametller & Pinto, 2002). In addition, middle- and high-school students tend to use their conceptualizations of energy to interpret energy conservation ideas (Brook & Driver, 1986; Kesidou & Duit, 1993; Solomon, 1985). For example, some students interpret the idea that "energy is not created or destroyed" to mean that energy is stored up in the system and can even be released again in its original form (Solomon, 1985). Or, students may believe that no energy remains at the end of a process, but may say that "energy is not lost" because an effect was caused during the process (for example, a weight was lifted) (Duit & Haeussler, 1994). Although teaching approaches which accommodate students' difficulties about energy appear to be more successful than traditional science instruction, the main deficiencies outlined above remain despite these approaches (Brook & Driver, 1986; Brook & Wells, 1988).

See **ENERGY RESOURCES** for additional research.

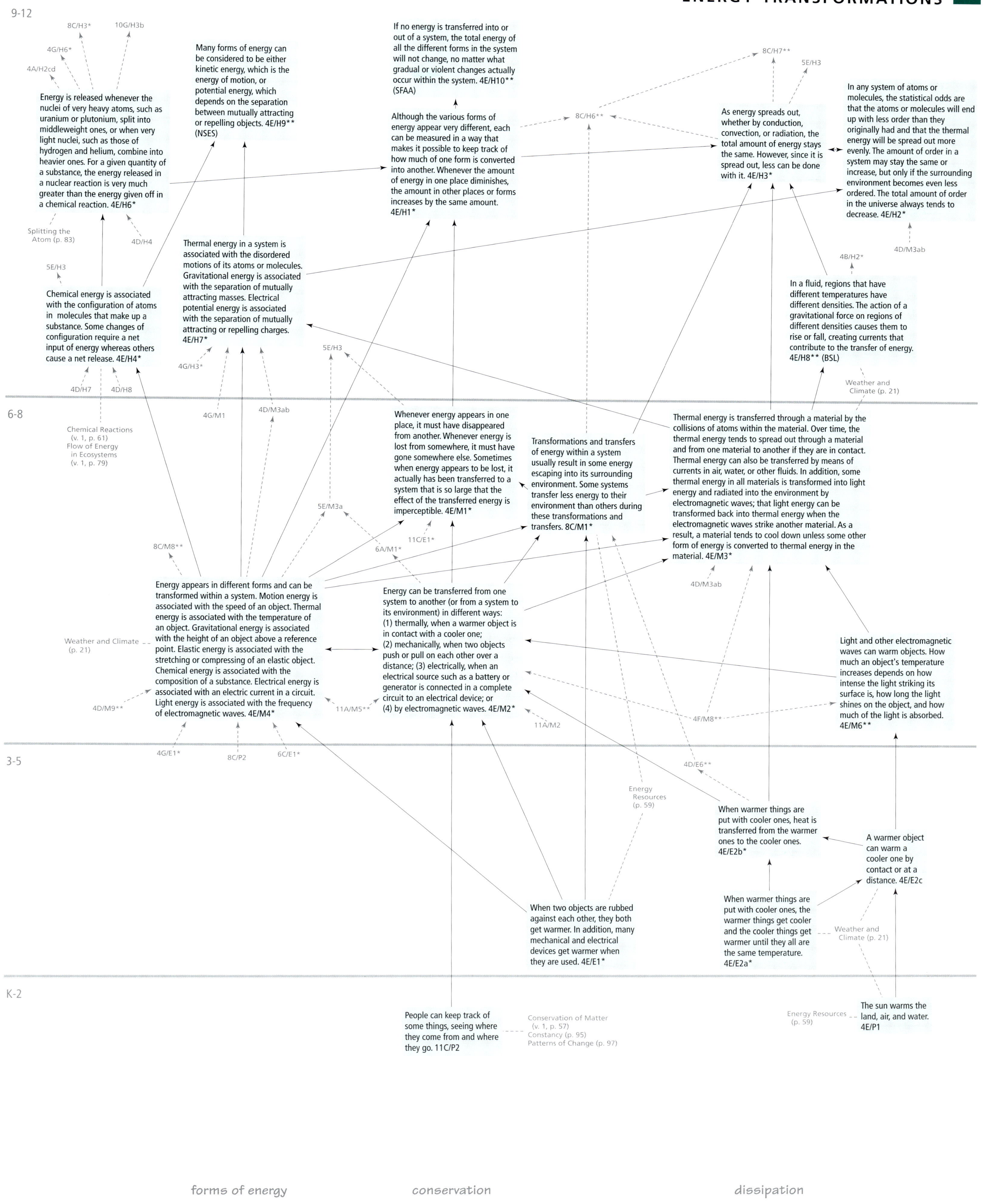
9-12
8C/H3*
10G/H3b
4G/H6*
4A/H2cd
Energy is released whenever the nuclei of very heavy atoms, such as uranium or plutonium, split into middleweight ones, or when very light nuclei, such as those of hydrogen and helium, combine into heavier ones. For a given quantity of a substance, the energy released in a nuclear reaction is very much greater than the energy given off in a chemical reaction. 4E/H6*
Splitting the Atom (p. 83)
4D/H4
Many forms of energy can be considered to be either kinetic energy, which is the energy of motion, or potential energy, which depends on the separation between mutually attracting or repelling objects. 4E/H9** (NSES)
If no energy is transferred into or out of a system, the total energy of all the different forms in the system will not change, no matter what gradual or violent changes actually occur within the system. 4E/H10** (SFAA)
Although the various forms of energy appear very different, each can be measured in a way that makes it possible to keep track of how much of one form is converted into another. Whenever the amount of energy in one place diminishes, the amount in other places or forms increases by the same amount. 4E/H1*
8C/H6**
8C/H7**
5E/H3
As energy spreads out, whether by conduction, convection, or radiation, the total amount of energy stays the same. However, since it is spread out, less can be done with it. 4E/H3*
In any system of atoms or molecules, the statistical odds are that the atoms or molecules will end up with less order than they originally had and that the thermal energy will be spread out more evenly. The amount of order in a system may stay the same or increase, but only if the surrounding environment becomes even less ordered. The total amount of order in the universe always tends to decrease. 4E/H2*
4D/M3ab
4B/H2*
5E/H3
Chemical energy is associated with the configuration of atoms in molecules that make up a substance. Some changes of configuration require a net input of energy whereas others cause a net release. 4E/H4*
4D/H7
4D/H8
Thermal energy in a system is associated with the disordered motions of its atoms or molecules. Gravitational energy is associated with the separation of mutually attracting masses. Electrical potential energy is associated with the separation of mutually attracting or repelling charges. 4E/H7*
4G/H3*
5E/H3
In a fluid, regions that have different temperatures have different densities. The action of a gravitational force on regions of different densities causes them to rise or fall, creating currents that contribute to the transfer of energy. 4E/H8** (BSL)
Weather and Climate (p. 21)
6-8
Chemical Reactions (v. 1, p. 61)
Flow of Energy in Ecosystems (v. 1, p. 79)
4G/M1
4D/M3ab
Whenever energy appears in one place, it must have disappeared from another. Whenever energy is lost from somewhere, it must have gone somewhere else. Sometimes when energy appears to be lost, it actually has been transferred to a system that is so large that the effect of the transferred energy is imperceptible. 4E/M1*
Transformations and transfers of energy within a system usually result in some energy escaping into its surrounding environment. Some systems transfer less energy to their environment than others during these transformations and transfers. 8C/M1*
Thermal energy is transferred through a material by the collisions of atoms within the material. Over time, the thermal energy tends to spread out through a material and from one material to another if they are in contact. Thermal energy can also be transferred by means of currents in air, water, or other fluids. In addition, some thermal energy in all materials is transformed into light energy and radiated into the environment by electromagnetic waves; that light energy can be transformed back into thermal energy when the electromagnetic waves strike another material. As a result, a material tends to cool down unless some other form of energy is converted to thermal energy in the material. 4E/M3*
5E/M3a
8C/M8**
11C/E1*
6A/M1*
4D/M3ab
Energy appears in different forms and can be transformed within a system. Motion energy is associated with the speed of an object. Thermal energy is associated with the temperature of an object. Gravitational energy is associated with the height of an object above a reference point. Elastic energy is associated with the stretching or compressing of an elastic object. Chemical energy is associated with the composition of a substance. Electrical energy is associated with an electric current in a circuit. Light energy is associated with the frequency of electromagnetic waves. 4E/M4*
Weather and Climate (p. 21)
4D/M9**
Energy can be transferred from one system to another (or from a system to its environment) in different ways:
(1) thermally, when a warmer object is in contact with a cooler one;
(2) mechanically, when two objects push or pull on each other over a distance; (3) electrically, when an electrical source such as a battery or generator is connected in a complete circuit to an electrical device; or
(4) by electromagnetic waves. 4E/M2*
11A/M5**
11A/M2
4F/M8**
Light and other electromagnetic waves can warm objects. How much an object's temperature increases depends on how intense the light striking its surface is, how long the light shines on the object, and how much of the light is absorbed. 4E/M6**
4G/E1*
8C/P2
6C/E1*
3-5
4D/E6**
Energy Resources (p. 59)
When warmer things are put with cooler ones, heat is transferred from the warmer ones to the cooler ones. 4E/E2b*
A warmer object can warm a cooler one by contact or at a distance. 4E/E2c
When two objects are rubbed against each other, they both get warmer. In addition, many mechanical and electrical devices get warmer when they are used. 4E/E1*
When warmer things are put with cooler ones, the warmer things get cooler and the cooler things get warmer until they all are the same temperature. 4E/E2a*
Weather and Climate (p. 21)
K-2
People can keep track of some things, seeing where they come from and where they go. 11C/P2
Conservation of Matter (v. 1, p. 57)
Constancy (p. 95)
Patterns of Change (p. 97)
Energy Resources (p. 59)
The sun warms the land, air, and water. 4E/P1
forms of energy
conservation
dissipation

# ELECTRICITY AND MAGNETISM (4G)

With the exception of gravity, electromagnetic forces are responsible for nearly all familiar phenomena—properties of materials that result from the interactions between atoms that make them up; chemical reactions; electric motors, generators, and transformers; and electromagnetic radiation.

The map is organized around four strands—*effects at the atomic level, electric charge, electric currents,* and *electromagnetism*. In the elementary grades, the focus is on characterizing various "pushes and pulls." In middle school, the focus is on the effect of forces on objects. High school benchmarks explain the origin of electric and nuclear forces in terms of the structure of the atom, characterize current electricity in terms of the motion of electrons, and relate electric to magnetic forces.

## NOTES

In grades 3-5, the *electric charge* strand focuses on observations of electrical interactions, but in grades 6-8 a new benchmark that begins "A charged object..." signals that students are to understand the concept of "charge" as the property acquired by the interacting objects and to recognize that there are two ways in which an object can be charged. It should be noted that the phrase "a charged object can be charged in one of two ways" refers to the final state of the object, not the method used to get it to that state. By grades 9-12, the study of electric charge is linked to the study of the structure of atoms; protons and electrons are described as electrically charged particles.

Starting in the upper elementary years, an important notion to convey is that forces—electric, magnetic, and gravitational—act at a distance. This is depicted in the map by the double-headed arrows between benchmarks 4G/E2, 4G/E3 and benchmark 4G/E1 (which focuses on gravity and is included here as an important off-map connection). In high school, benchmark 4G/H2a compares the strength and effect of electric and gravitational interactions at the atomic and larger scales.

## RESEARCH IN *BENCHMARKS*

Before instruction, many elementary- and middle-school students are not aware of the bipolarity of batteries and light bulbs; do not recognize the need for a complete circuit to make a bulb light; and do not succeed in making a lamp light when given a battery and a number of connecting wires (Shipstone, 1985; Arnold & Millar, 1987; Borges & Gilbert, 1999). However, many high-school and university students also fail at this task (Shipstone, 1985; Arons, 1997). This suggests that they also do not understand or cannot apply the concept of a complete circuit (McDermott & Schaffer, 1992). Teaching sequences that take account of students' ideas can help middle-school students make progress in this area (Arnold & Millar, 1987). Students of all ages have difficulty reasoning that all parts of a circuit are interrelated and influence each other. Instead, they think of circuits in terms of electric current traveling around the circuit meeting each component in turn. They think of a change in the circuit affecting only those components that come after the change. This "sequential" reasoning underlies many problems that students have in understanding electric circuits and is highly resistant to change (Closset, 1983; Driver et al., 1994; Shipstone et al., 1988).

Students tend to start instruction with one concept for electricity in electric circuits which has the properties of movement, storability, and consumability and which students label "current," "energy," or "electricity." (Shipstone, 1985; Driver et al., 1994). Even after instruction, many students of all ages do not differentiate between electric current and electric energy. They also tend to think that the battery is the source of the current and that the circuit is initially empty of the stuff that flows through the wires (Tiberghien, 1983; Heller & Finley, 1992; McDermott & Schaffer, 1992). Many students after instruction believe that a battery releases the same amount of current regardless of the circuit to which it is attached, that the fixed current flows out of the battery and diminishes every time it goes through a circuit element that uses up the current, so that there is less current at the end of the circuit (Shipstone, 1985; McDermott & Schaffer, 1992). These beliefs are highly resistant to change (Gauld, 1989; Schaffer & McDermott, 1992; Shipstone et al., 1988). Identifying energy as the quantity that is dissipated can help students reconcile their intuitive belief that something is used up in circuits with the formal knowledge that electric current is conserved (Licht, 1991; Shaffer & McDermott, 1992).

Little is known about students' reasoning about the microscopic mechanisms that underlie electric current and their interpretation in terms of electrostatic entities. After instruction, high-school students may not be inclined to or, when prompted, may have difficulties relating macroscopic parameters (such as electric current) to microscopic processes and electrostatic interactions (such as forces on charged electrons) (Eylon & Ganiel, 1990). Students may think that of the battery as the only source of electrons which move in the circuit, i.e., the battery releases electrons into the wires which play no active role; they may also think of electrons moving through a circuit as single unconnected particles moving around (Borges & Gilbert, 1999; Eylon & Ganiel, 1990).

Elementary-school students are usually aware of the behavior of magnets but may not explain the behavior in terms of forces (i.e., they may think of a magnet sticking to or moving towards another magnet but may not recognize this as the effect of a pull or force) (Driver et al., 1994a). Students of all ages may think of gravity and magnetism interchangeably. They may refer to magnetism as a "type of gravity," but they may also explain gravity in terms of the earth acting like a magnet on objects. Students may think that magnets do not work in a place where there is no air, just like they think about gravity (Arons, 1997; Driver et al., 1994a). Students of all ages may also confuse electrostatic and magnetic effects (Arons, 1997). For example, they may predict that north magnetic poles repel positively charged objects (Maloney, 1985).

Students do not readily recognize the magnetic effect of an electric current. Some think of the wire, rather than the electric current as being the cause of the magnetic effect. Students may think that insulation around the wire prevents the existence of magnetic forces when current flows (Driver et al., 1994a).

9-12
The nuclear forces that hold the protons and neutrons in the nucleus of an atom together are much stronger than the electric forces between the protons and electrons of the atom. That is why much greater amounts of energy are released from nuclear reactions than from chemical reactions. 4G/H6*
4D/H3*
4E/H6*
Electric forces acting within and between atoms are vastly stronger than the gravitational forces acting between the atoms. At larger scales, gravitational forces accumulate to produce a large and noticable effect whereas electric forces tend to cancel each other out. 4G/H2a*
4G/H1
At the atomic level, electric forces between electrons and protons in atoms hold molecules together and thus are involved in all chemical reactions. 4G/H2b*
4D/H8
4D/H2*
4D/M1cd*
Chemical Reactions (v. 1, p. 61)
Atoms and Molecules (v. 1, p. 55)
Splitting the Atom (p. 83)
Electric forces hold solid and liquid materials together and act between objects when they are in contact—as in sticking or sliding friction. 4G/H2c
Most materials have equal numbers of protons and electrons and are therefore electrically neutral. In most cases, a material acquires a negative charge by gaining electrons and acquires a positive charge by losing electrons. Even a tiny imbalance in the number of protons and electrons in an object can produce noticeable electric forces on other objects. 4G/H3*
4E/H6*
Atoms are made of a positively charged nucleus surrounded by negatively charged electrons. The nucleus is a tiny fraction of the volume of an atom but makes up almost all of its mass. The nucleus is composed of protons and neutrons which have roughly the same mass, but differ in that protons are positively charged while neutrons have no electric charge. 4D/H1*
8B/H4*
Semiconducting materials differ greatly in how well they conduct electrons depending on the exact composition of the material. 4G/H4d*
At very low temperatures, some materials become superconductors and offer no resistance to the flow of electrons. 4G/H4c*
Materials Science (p. 55)
In many conducting materials, such as metals, some of the electrons are not firmly held by the nuclei of the atoms that make up the material. In these materials, applied electric forces can cause the electrons to move through the material, producing an electric current. In insulating materials, such as glass, the electrons are held more firmly, making it nearly impossible to produce an electric current in those materials. 4G/H4ab*
The motion of electrons is far more affected by electrical forces than protons are because electrons are much less massive and are outside of the nucleus. 4G/H8** (BSL)
The change in motion (direction or speed) of an object is proportional to the applied force and inversely proportional to the mass. 4F/H1*
12C/H3*
Electric currents in the earth's interior give the earth an extensive magnetic field, which we detect from the orientation of compass needles. 4G/H7** (SFAA)
4C/M1ab
The interplay of electric and magnetic forces is the basis for many modern technologies including electric motors, generators, and devices that produce or receive electromagnetic waves. 4G/H5c*
Magnetic forces are very closely related to electric forces and are thought of as different aspects of a single electromagnetic force. Moving electrically charged objects produces magnetic forces and moving magnets produces electric forces. 4G/H5ab*
When electrically charged objects undergo a change in motion, they produce electromagnetic waves around them. 4F/H3a*
4F/M8**
4F/M4
6-8
4D/M1a
Laws of Motion (v. 1, p. 63)
Classical Mechanics (p. 75)
4D/M6a
8C/M4*
Electrical circuits require a complete loop through which an electrical current can pass. 4G/M4** (NSES)
Materials vary in how they respond to electric currents, magnetic forces, and visible light or other electromagnetic waves. 4D/M9**
Using Tools and Devices (p. 109)
Materials Science (p. 55)
A charged object can be charged in one of two ways, which we call either positively charged or negatively charged. Two objects that are charged in the same manner exert a force of repulsion on each other, while oppositely charged objects exert a force of attraction on each other. 4G/M5** (BSL)
An unbalanced force acting on an object changes its speed or direction of motion, or both. 4F/M3a
3-5
Gravity (v. 1, p. 43)
Laws of Motion (v. 1, p. 63)
4F/E1bc
Changes in speed or direction of motion are caused by forces. 4F/E1a
Without touching them, an object that has been electrically charged pulls on all other uncharged objects and may either push or pull other charged objects. 4G/E3*
4G/E1*
Without touching them, a magnet pulls on all things made of iron and either pushes or pulls on other magnets. 4G/E2
K-2
4F/P2
The way to change how something is moving is to give it a push or a pull. 4F/P2
Gravity (v. 1, p. 43)
Laws of Motion (v. 1, p. 63)
Magnets can be used to make some things move without being touched. 4G/P2
effects at the atomic level
electric charge
electric currents
electromagnetism

Henri Rousseau *Tropical Forest with Monkeys,* 1910

*People have long been curious about living things—how many different species there are, what they are like, where they live, how they relate to each other, and how they behave. Scientists seek to answer these questions and many more about the organisms that inhabit the earth. In particular, they try to develop the concepts, principles, and theories that enable people to understand the living environment better.*

*Living organisms are made of the same components as all other matter, involve the same kind of transformations of energy, and move using the same basic kinds of forces. Thus, all of the physical principles discussed in Chapter 4,* THE PHYSICAL SETTING, *apply to life as well as to stars, raindrops, and television sets. But living organisms also have characteristics that can be understood best through the application of other principles.*

SCIENCE FOR ALL AMERICANS

# CHAPTER 5

# THE LIVING ENVIRONMENT



Understanding the living environment involves not only understanding various levels of biological organization—molecules, cells, tissues and organs, organisms, populations, and ecosystems—but also understanding interactions within and across levels and how these interactions can change over time. Because life processes, such as inheritance and the transformation of matter and energy, involve multiple levels of biological organization, and because those levels interact with one another, any attempt to organize life science by a linear set of topics, such as those represented in the maps for this chapter, is somewhat arbitrary. Nevertheless, each map offers a unique context within which central ideas having to do with the structure, function, and patterns of change in living systems are presented. Chapter 6: THE HUMAN ORGANISM augments many of these ideas in the context of human beings.

The list on the left of this page shows the maps for this chapter and indicates the volume of *Atlas of Science Literacy* in which each appears. Those in green typeface are in *Volume 2* and those in gray are in *Volume 1*.

# DIVERSITY OF LIFE (5A)

There is an enormous variety of types of organisms, including humans, that inhabit the earth at any one time—some very similar to one another and some very different. Appreciating that diversity, its underlying unity, and the potential consequences of a reduction in diversity are important for science literacy.

The map is organized around three strands—*classification, similarities and differences,* and *diversity and survival*. The progression of understanding begins in the elementary grades concretely with an emphasis on similarities between living organisms and on grouping organisms based on those similarities. In middle school the emphasis shifts to an explicit definition of species; to similarities based on internal features, including cells; and to the diverse and changing environments inhabited by a wide variety of organisms. By high school, students are better able to appreciate the underlying molecular unity of living organisms and the potential consequences of reducing the diversity of life.

Ideas in this map are closely linked to ideas about heredity, cells, and evolution, as shown by the numerous connections on the map. The accepted mechanisms for ensuring the diversity of life (natural selection operating in ever changing environments) and the underlying similarity within diversity (descent from common ancestors) are treated in the **NATURAL SELECTION** and **BIOLOGICAL EVOLUTION** maps in *Atlas 1*. The complex relationships among species that contribute to ecosystem stability are treated in the **INTERDEPENDENCE OF LIFE** map.

### NOTES

The *similarities and differences* strand presents a progressively more sophisticated perspective on the features of organisms that connect or differentiate them: from external features and behavior patterns, to internal structures, to cellular activity and patterns of development, and to molecular structure. Several benchmarks in this strand also contribute to the evidence for evolution in the **BIOLOGICAL EVOLUTION** map.

The point of the *classification* strand is to develop understanding about the purpose and role of classification, not about the details of standard classification systems.

The *diversity and survival* strand develops the idea that a great diversity of life increases the chance of survival of some kinds of living things. The grades 9-12 benchmark "The continuing operation of natural selection..." addresses the question of how life became diverse and is supported by benchmarks on diversity of environments and the rise of new heritable characteristics. All three benchmarks are also part of the **NATURAL SELECTION** map, which includes a more extensive progression on explaining the diversity of life.

### RESEARCH IN *BENCHMARKS*

Research has focused on what students understand about the living environment at isolated points in time or on how this understanding evolves naturally in students. Research on instructional interventions that improve students' understanding is limited. Reviews of research can be found in Carey (1985), Good et al. (1993), and Mintzes et al. (1991).

Some research indicates that in 2nd grade there is a shift in children's understanding of organisms from representations based on perceptual and behavioral features to representations in which central principles of biological theory are most important. Children at this age can begin to understand that animals of the same species have similar internal parts and offspring (Keil, 1989). When asked to group certain organisms, lower elementary-school students form groups of different status—for example, organisms that are able to fly and organisms that fight each other. Upper elementary-school students tend to use a number of mutually exclusive groups rather than a hierarchy of groups. Some groups are based on observable features; others on concepts. By middle school, students can group organisms hierarchically when asked to do so, whereas high-school students use hierarchical taxonomies without prompting (Leach et al., 1992).

Elementary- and middle-school students hold a much more restricted meaning than biologists for the word "animal" (Mintzes et al., 1991). For example, most students list only vertebrates as animals. Elementary- and middle-school students use such criteria as number of legs, body covering, and habitat to decide whether things are animals. High-school students frequently use attributes that are common to both plants and animals (e.g., reproduction and respiration) as criteria (Trowbridge & Mintzes, 1985). Because upper elementary-school students tend not to use hierarchical classification, they may have difficulty understanding that an organism can be classified as both a bird and an animal (Bell, 1981). Elementary- and middle-school students also hold a much more restricted meaning than biologists do for the word "plant." Students often do not recognize that trees, vegetables, and grass are all plants (Osborne & Freyberg, 1985).

Elementary- and middle-school students typically use criteria such as "movement," "breath," "reproduction," and "death" to decide whether things are alive. Thus, some believe fire, clouds, and the sun are alive, but others think plants and certain animals are nonliving (Bell & Freyberg, 1985; Leach et al., 1992). High-school and college students also mainly use obvious criteria (e.g., "movement," "growth") to distinguish between "living" and "nonliving" and rarely mention structural criteria ("cells") or biochemical characteristics ("DNA") (Brumby, 1982; Leach et al., 1992).

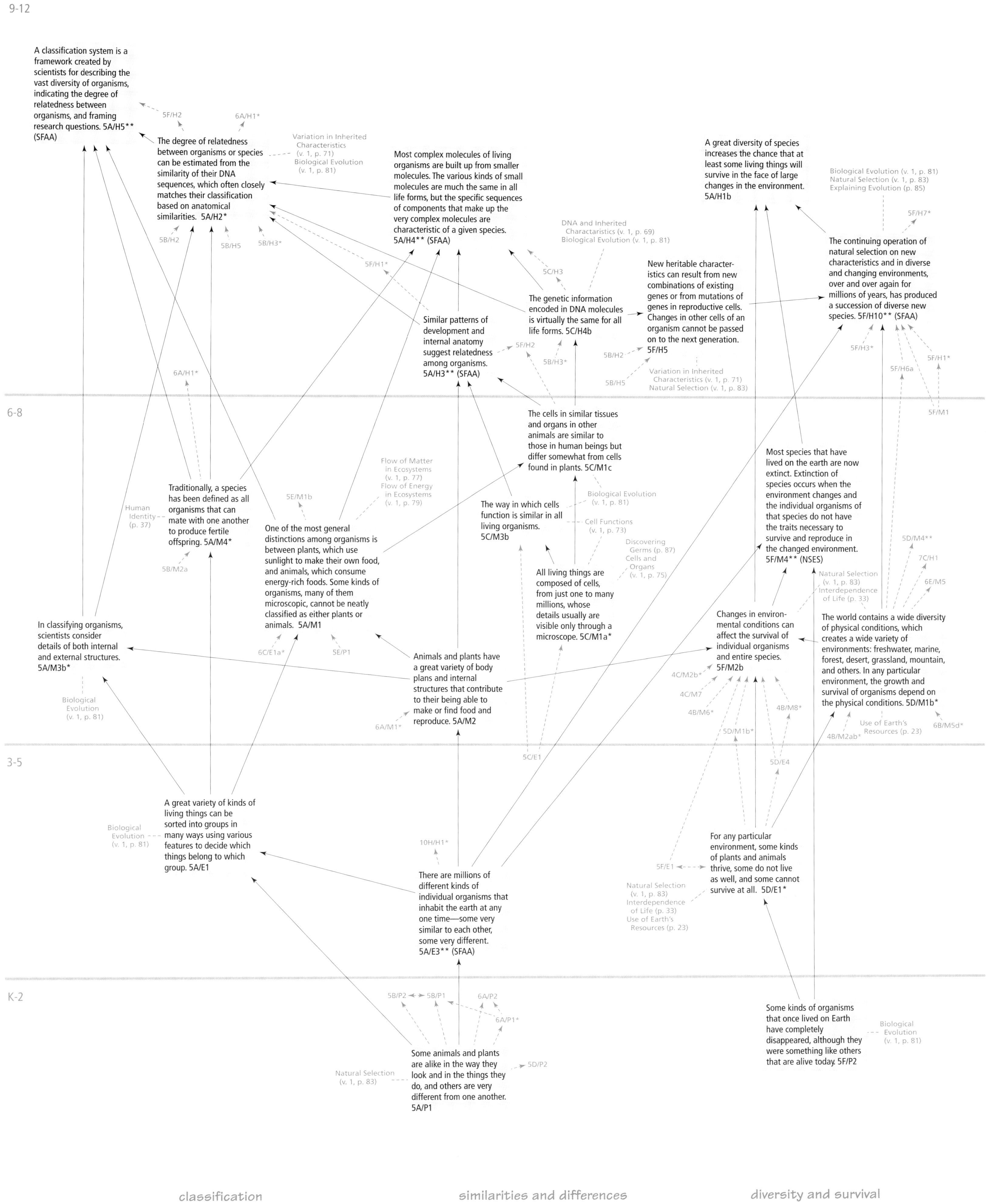
9-12
A classification system is a framework created by scientists for describing the vast diversity of organisms, indicating the degree of relatedness between organisms, and framing research questions. 5A/H5** (SFAA)
5F/H2
6A/H1*
The degree of relatedness between organisms or species can be estimated from the similarity of their DNA sequences, which often closely matches their classification based on anatomical similarities. 5A/H2*
Variation in Inherited Characteristics (v. 1, p. 71)
Biological Evolution (v. 1, p. 81)
5B/H2
5B/H5
5B/H3*
Most complex molecules of living organisms are built up from smaller molecules. The various kinds of small molecules are much the same in all life forms, but the specific sequences of components that make up the very complex molecules are characteristic of a given species. 5A/H4** (SFAA)
5F/H1*
A great diversity of species increases the chance that at least some living things will survive in the face of large changes in the environment. 5A/H1b
Biological Evolution (v. 1, p. 81)
Natural Selection (v. 1, p. 83)
Explaining Evolution (p. 85)
5F/H7*
DNA and Inherited Charactaristics (v. 1, p. 69)
Biological Evolution (v. 1, p. 81)
5C/H3
The continuing operation of natural selection on new characteristics and in diverse and changing environments, over and over again for millions of years, has produced a succession of diverse new species. 5F/H10** (SFAA)
New heritable characteristics can result from new combinations of existing genes or from mutations of genes in reproductive cells. Changes in other cells of an organism cannot be passed on to the next generation. 5F/H5
The genetic information encoded in DNA molecules is virtually the same for all life forms. 5C/H4b
Similar patterns of development and internal anatomy suggest relatedness among organisms. 5A/H3** (SFAA)
5F/H2
5B/H3*
5B/H2
5B/H5
Variation in Inherited Characteristics (v. 1, p. 71)
Natural Selection (v. 1, p. 83)
5F/H3*
5F/H6a
5F/H1*
6A/H1*
6-8
5F/M1
The cells in similar tissues and organs in other animals are similar to those in human beings but differ somewhat from cells found in plants. 5C/M1c
Most species that have lived on the earth are now extinct. Extinction of species occurs when the environment changes and the individual organisms of that species do not have the traits necessary to survive and reproduce in the changed environment. 5F/M4** (NSES)
Flow of Matter in Ecosystems (v. 1, p. 77)
Flow of Energy in Ecosystems (v. 1, p. 79)
5E/M1b
Traditionally, a species has been defined as all organisms that can mate with one another to produce fertile offspring. 5A/M4*
Human Identity (p. 37)
5B/M2a
One of the most general distinctions among organisms is between plants, which use sunlight to make their own food, and animals, which consume energy-rich foods. Some kinds of organisms, many of them microscopic, cannot be neatly classified as either plants or animals. 5A/M1
The way in which cells function is similar in all living organisms. 5C/M3b
Biological Evolution (v. 1, p. 81)
Cell Functions (v. 1, p. 73)
Discovering Germs (p. 87)
Cells and Organs (v. 1, p. 75)
All living things are composed of cells, from just one to many millions, whose details usually are visible only through a microscope. 5C/M1a*
5D/M4**
7C/H1
6E/M5
Natural Selection (v. 1, p. 83)
Interdependence of Life (p. 33)
In classifying organisms, scientists consider details of both internal and external structures. 5A/M3b*
6C/E1a*
5E/P1
Changes in environmental conditions can affect the survival of individual organisms and entire species. 5F/M2b
The world contains a wide diversity of physical conditions, which creates a wide variety of environments: freshwater, marine, forest, desert, grassland, mountain, and others. In any particular environment, the growth and survival of organisms depend on the physical conditions. 5D/M1b*
Biological Evolution (v. 1, p. 81)
Animals and plants have a great variety of body plans and internal structures that contribute to their being able to make or find food and reproduce. 5A/M2
6A/M1*
4C/M2b*
4C/M7
4B/M6*
4B/M8*
5D/M1b*
Use of Earth's Resources (p. 23)
4B/M2ab*
6B/M5d*
3-5
5C/E1
5D/E4
A great variety of kinds of living things can be sorted into groups in many ways using various features to decide which things belong to which group. 5A/E1
Biological Evolution (v. 1, p. 81)
10H/H1*
There are millions of different kinds of individual organisms that inhabit the earth at any one time—some very similar to each other, some very different. 5A/E3** (SFAA)
For any particular environment, some kinds of plants and animals thrive, some do not live as well, and some cannot survive at all. 5D/E1*
5F/E1
Natural Selection (v. 1, p. 83)
Interdependence of Life (p. 33)
Use of Earth's Resources (p. 23)
K-2
5B/P2
5B/P1
6A/P2
6A/P1*
Some animals and plants are alike in the way they look and in the things they do, and others are very different from one another. 5A/P1
Natural Selection (v. 1, p. 83)
5D/P2
Some kinds of organisms that once lived on Earth have completely disappeared, although they were something like others that are alive today. 5F/P2
Biological Evolution (v. 1, p. 81)
classification
similarities and differences
diversity and survival

## THE LIVING ENVIRONMENT

# INTERDEPENDENCE OF LIFE (5D)

Every species is linked, directly or indirectly, with a multitude of others in an ecosystem. Ecosystems are shaped by both the nonliving environment and by its inhabitants, including humans. Hence, benchmarks on this map are closely related to benchmarks on the **FLOW OF MATTER IN ECOSYSTEMS** map in *Atlas 1* and to benchmarks on the **USE OF EARTH'S RESOURCES** map.

The map is organized around four strands—*interactions among organisms, dynamic nature of ecosystems, dependence of organisms on their environment,* and *human impact*. The learning progression begins in the elementary grades with an emphasis on the needs of organisms and how they are met in different environments. In middle school, the emphasis is on understanding how organisms (including humans) interact with one another and with the environment in a wide variety of ecosystems. In high school, students' knowledge about the interdependence of organisms in ecosystems and the environment is linked to more abstract ideas about stability and change in systems.

### NOTES

Students may not regard food as a scarce resource for animals and, hence, may not consider competition among species for food resources. Therefore, the grades 6-8 benchmark "In all environments, organisms with similar needs…" was modified to include the idea that resources for which animals compete are limited. A new benchmark (5D/M3) was added to the map to extend students' understanding of the implications of finite resources for populations of organisms.

Students' knowledge of the variety of environments and changes in environmental conditions on Earth needs to be integrated with their growing understanding of earth science, in particular with benchmarks related to climate in the **WEATHER AND CLIMATE** map. As students become more familiar with the characteristics of systems in general, they can begin to recognize some of those characteristics—such as interdependence of parts, stability, and change—as they appear in ecosystems.

Ethical choices and the implications of various uses of the environment are not explicitly addressed in this map. However, the map can provide contexts for learning about the effects of science on society, ethics in research, and the incompleteness of scientific answers as presented on the **SCIENCE AND SOCIETY** map and the **SCIENTIFIC WORLD VIEW** map.

### RESEARCH IN *BENCHMARKS*

Lower elementary-school students can understand simple food links involving two organisms. Yet they often think of organisms as independent of each other but dependent on people to supply them with food and shelter. Upper elementary-school students may not believe food is a scarce resource in ecosystems, thinking that organisms can change their food at will according to the availability of particular sources (Leach et al., 1992). Students of all ages think that some populations of organisms are numerous in order to fulfill a demand for food by another population (Leach et al., 1992).

Middle-school and high-school students may believe that organisms are able to effect changes in bodily structure to exploit particular habitats or that they respond to a changed environment by seeking a more favorable environment (Jungwirth, 1975; Clough & Wood-Robinson, 1985a). It has been suggested that the language about adaptation used by teachers or textbooks to make biology more accessible to students may cause or reinforce these beliefs (Jungwirth, 1975).

Some middle-school students think dead organisms simply rot away. They do not realize that the matter from the dead organism is converted into other materials in the environment. Some middle-school students see decay as a gradual, inevitable consequence of time without need of decomposing agents (Smith & Anderson, 1986). Some high-school students believe that matter is conserved during decay, but do not know where it goes (Leach et al., 1992).

**THE FAR SIDE®** By GARY LARSON

"Mind if we check the ears?"

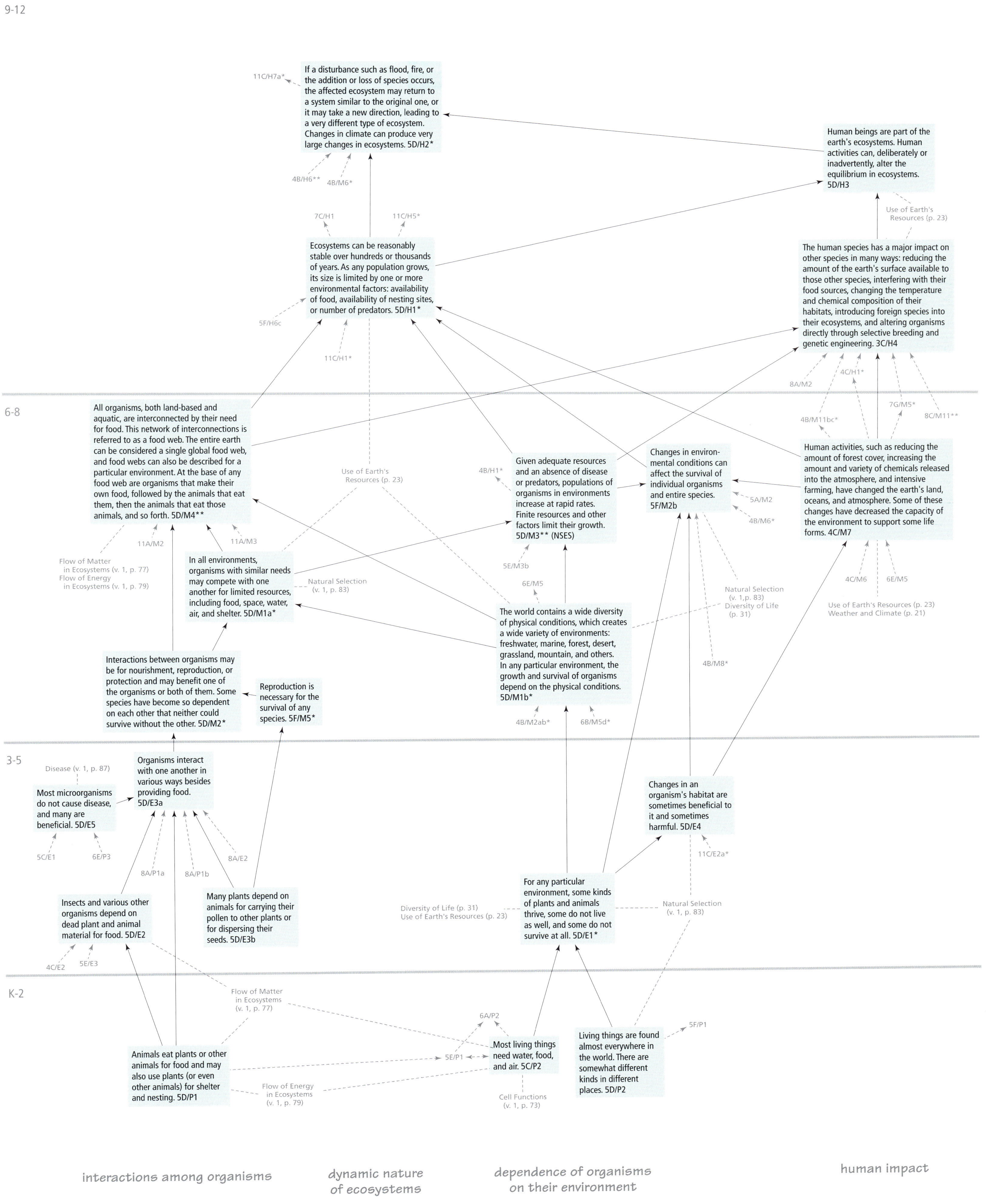
9-12
If a disturbance such as flood, fire, or the addition or loss of species occurs, the affected ecosystem may return to a system similar to the original one, or it may take a new direction, leading to a very different type of ecosystem. Changes in climate can produce very large changes in ecosystems. 5D/H2*
11C/H7a*
4B/H6**
4B/M6*
Human beings are part of the earth's ecosystems. Human activities can, deliberately or inadvertently, alter the equilibrium in ecosystems. 5D/H3
Use of Earth's Resources (p. 23)
7C/H1
11C/H5*
Ecosystems can be reasonably stable over hundreds or thousands of years. As any population grows, its size is limited by one or more environmental factors: availability of food, availability of nesting sites, or number of predators. 5D/H1*
5F/H6c
11C/H1*
The human species has a major impact on other species in many ways: reducing the amount of the earth's surface available to those other species, interfering with their food sources, changing the temperature and chemical composition of their habitats, introducing foreign species into their ecosystems, and altering organisms directly through selective breeding and genetic engineering. 3C/H4
8A/M2
4C/H1*
4B/M11bc*
7G/M5*
8C/M11**
6-8
All organisms, both land-based and aquatic, are interconnected by their need for food. This network of interconnections is referred to as a food web. The entire earth can be considered a single global food web, and food webs can also be described for a particular environment. At the base of any food web are organisms that make their own food, followed by the animals that eat them, then the animals that eat those animals, and so forth. 5D/M4**
Use of Earth's Resources (p. 23)
4B/H1*
Given adequate resources and an absence of disease or predators, populations of organisms in environments increase at rapid rates. Finite resources and other factors limit their growth. 5D/M3** (NSES)
Changes in environmental conditions can affect the survival of individual organisms and entire species. 5F/M2b
5A/M2
4B/M6*
Human activities, such as reducing the amount of forest cover, increasing the amount and variety of chemicals released into the atmosphere, and intensive farming, have changed the earth's land, oceans, and atmosphere. Some of these changes have decreased the capacity of the environment to support some life forms. 4C/M7
11A/M2
11A/M3
Flow of Matter in Ecosystems (v. 1, p. 77)
Flow of Energy in Ecosystems (v. 1, p. 79)
In all environments, organisms with similar needs may compete with one another for limited resources, including food, space, water, air, and shelter. 5D/M1a*
Natural Selection (v. 1, p. 83)
5E/M3b
6E/M5
The world contains a wide diversity of physical conditions, which creates a wide variety of environments: freshwater, marine, forest, desert, grassland, mountain, and others. In any particular environment, the growth and survival of organisms depend on the physical conditions. 5D/M1b*
Natural Selection (v. 1, p. 83)
Diversity of Life (p. 31)
4C/M6
6E/M5
Use of Earth's Resources (p. 23)
Weather and Climate (p. 21)
4B/M8*
Interactions between organisms may be for nourishment, reproduction, or protection and may benefit one of the organisms or both of them. Some species have become so dependent on each other that neither could survive without the other. 5D/M2*
Reproduction is necessary for the survival of any species. 5F/M5*
4B/M2ab*
6B/M5d*
3-5
Disease (v. 1, p. 87)
Most microorganisms do not cause disease, and many are beneficial. 5D/E5
Organisms interact with one another in various ways besides providing food. 5D/E3a
Changes in an organism's habitat are sometimes beneficial to it and sometimes harmful. 5D/E4
5C/E1
6E/P3
8A/P1a
8A/P1b
8A/E2
11C/E2a*
Insects and various other organisms depend on dead plant and animal material for food. 5D/E2
Many plants depend on animals for carrying their pollen to other plants or for dispersing their seeds. 5D/E3b
Diversity of Life (p. 31)
Use of Earth's Resources (p. 23)
For any particular environment, some kinds of plants and animals thrive, some do not live as well, and some do not survive at all. 5D/E1*
Natural Selection (v. 1, p. 83)
4C/E2
5E/E3
K-2
Flow of Matter in Ecosystems (v. 1, p. 77)
6A/P2
Animals eat plants or other animals for food and may also use plants (or even other animals) for shelter and nesting. 5D/P1
5E/P1
Most living things need water, food, and air. 5C/P2
Living things are found almost everywhere in the world. There are somewhat different kinds in different places. 5D/P2
5F/P1
Flow of Energy in Ecosystems (v. 1, p. 79)
Cell Functions (v. 1, p. 73)
interactions among organisms
dynamic nature of ecosystems
dependence of organisms on their environment
human impact

Tao-chi *Among Peaks and Pines (Mt. Huang)*, ca. 1701

*As similar as we humans are in many ways to other species, we are unique among the earth's life forms in our ability to use language and thought. Having evolved a large and complex brain, our species has a facility to think, imagine, create, and learn from experience that far exceeds that of any other species. We have used this ability to create technologies and literary and artistic works on a vast scale, and to develop a scientific understanding of ourselves and the world.*

*We are also unique in our profound curiosity about ourselves: How are we put together physically? How were we formed? How do we relate biologically to other life forms and to our ancestors? How are we as individuals like or unlike other humans? How can we stay healthy? Much of the scientific endeavor focuses on such questions.*

SCIENCE FOR ALL AMERICANS

# CHAPTER 6



# THE HUMAN ORGANISM

Knowing about humans as a species is an important aspect of science literacy. Ideas mapped in this chapter relate to many of the ideas found in Chapter 5: THE LIVING ENVIRONMENT. Indeed, many of the characteristics of the human organism that are treated in the maps in this chapter are common to all mammals, or all animals, or all life forms. They are presented in a human context because that is easiest to learn for many students. Still, some features of life may be less readily recognizable in human beings because they are covered by layers of socialization and language. People sometimes become aware of their own characteristics only when they see them in other animals. Although *Science for All Americans* and *Benchmarks for Science Literacy* include recommendations for what students should know about the topic of learning, neither volume of the *Atlas* maps those ideas. Future editions will update those recommendations and map them.

The list on the left of this page shows the maps for this chapter and indicates the volume of *Atlas of Science Literacy* in which each appears. Those in green typeface are in *Volume 2* and those in gray are in *Volume 1*.

# HUMAN IDENTITY (6A)

The human species is biological, social, cultural, and technological. Humans are like other organisms in their chemical composition, basic cellular mechanisms, and more. However, humans have a much greater range of social behavior than other species and develop and use technology to more than make up for their deficiencies of speed, agility, stamina, vision, hearing, or the ability to withstand extremes of environmental conditions. Students should come to appreciate their kinship with other organisms, as well as their uniqueness.

The map is organized around four strands—*similarity to other species, physical attributes of the human species, behavioral attributes of the human species,* and *use of technology*. In the elementary grades the focus is on people's readily observable attributes and use of technology; in middle school the emphasis is on similarities and differences between humans and other species, paying particular attention to humans' ability to develop complex technologies; in high school the consideration of similarities and differences expands to include the molecular level and the human brain and its importance in producing knowledge and technology.

In addition to having close links to other maps in Chapter 6: THE HUMAN ORGANISM, this map is also relevant to maps in Chapter 7: HUMAN SOCIETY, particularly **HEREDITY AND EXPERIENCE SHAPE BEHAVIOR** and **CULTURE AFFECTS BEHAVIOR** in *Atlas 1*.

## NOTES

The *similarity to other species* strand focuses on the ideas that human beings are biologically very similar to other living organisms and have evolved from other organisms. The strand progresses from similarities in needs, to similarities in organ systems and cellular make up, to similarities in molecular structure and genetic code.

The *physical attributes of the human species* strand focuses on the idea that human beings have differences in appearance (and culture) but they are one species—distinctly different from other species. The 6-8 benchmark "Like other complex organisms..." draws from benchmarks on humans' unique technological abilities and wide range of social behavior, and provides a good summation of the central idea.

Learning defines a prominent way in which human beings are different from other species. The ability to transmit ideas and practices from one generation to the next has resulted in the enormous range of human behavior and distinguishes human beings from other species. In the early grades, the *behavioral attributes of the human species* strand focuses on learning. In high school, the ability to learn and the resulting range of human behavior is connected to the complexity of the human brain.

## RESEARCH IN *BENCHMARKS*

Several studies have investigated the spontaneous development of student conceptions about the human organism. A general pattern that emerged from this research is that until the age of 7, students have little knowledge about the human organism. By age 9 or 10, there is a marked increase in their knowledge (Carey, 1985).

Many U.S. adults who accept the idea that plants and animals have evolved reject the idea that humans have evolved. This appears to be related both to their religious beliefs and to their lack of understanding of the molecular basis of heredity (Miller, Scott, & Okamoto, 2006).

9-12

Fossil and molecular evidence supports the idea that human beings evolved from earlier species. 6A/H3** (BSL)

The similarity of humans in their cell chemistry and DNA sequences reinforces the idea that all humans are part of a single species. 6A/H1*

Cells and Organs (v. 1, p. 75)

The complexity of the human brain allows humans to create technological, literary, and artistic works on a vast scale, and to develop a scientific understanding of the world. 6B/H3*

Human Development (p. 39)

5F/H2

The degree of relatedness between organisms or species can be estimated from the similarity of their DNA sequences, which often closely match their classification based on anatomical similarities. 5A/H2*

5C/H4b

5C/H4a

6D/H3

The human ability to influence the course of history comes from its capacity for generating knowledge and developing new technologies—and for communicating ideas to others. 3C/H6** (BSL)

5F/H1*

Similar patterns of development and internal anatomy suggest relatedness among organisms. 5A/H3** (SFAA)

6-8

The cells in similar tissues and organs in other animals are similar to those in human beings but differ somewhat from cells found in plants. 5C/M1c

Like other complex organisms, people vary somewhat in size and shape, skin color, body proportions, body hair, facial features, muscle strength, handedness, and so on. But these differences are minor compared to the overall similarity of all humans, as demonstrated by the fact that people from anywhere in the world can reproduce with each other and donate blood or organs to one another. Humans are indeed a single species. Furthermore, as great as cultural differences between groups of people seem to be, people's complex languages, technologies, and arts unite them as a species distinct from others. 6A/M7** (SFAA)

Although social behaviors are affected by both genes and environmental factors, human beings are still able to invent, learn, and modify a wide variety of these behaviors. 6A/M4*

Technology is largely responsible for the great revolutions in agriculture, manufacturing, sanitation and medicine, warfare, transportation, information processing, and communications that have radically changed how people live and work. 3C/M4*

Technology and Science (p. 17)
Public Perception of Science (p. 105)
Discovering Germs (p. 87)
The Industrial Revolution (p. 89)
Interaction of Technology and Society (v. 1, p. 37)

5B/M2a

7A/M2*

6D/M1b

8A/M1

5A/M4*

7A/E3**

Basic Functions (p. 41)

Human beings use technology to match or exceed many of the abilities of other species. 6A/M5a*

Technologies having to do with food production, sanitation, and health care have dramatically changed how people live and work and have resulted in rapid increases in the human population. 6A/M6*

Written records and photographic and electronic devices enable human beings to share, compile, and use great amounts of information. 6A/M8** (BSL)

Like other animals, human beings have body systems for obtaining and deriving energy from food and for defense, reproduction, and the coordination of body functions. 6A/M1*

Interactions among the senses, nerves, and brain make possible the learning that enables human beings to predict, analyze, and respond to changes in their environment. 6C/M6*

Interaction of Technology and Society (v. 1, p. 37)
Influences on Social Change (v. 1, p. 101)

6C/M2

8A/M1

8F/M1*

8D/M2

3-5

6C/E4*

Diagnosis and Treatment of Mental Disorders (v. 1, p. 93)
Coping with Mental Distress (v. 1, p. 91)
Basic Functions (p. 41)

5B/E1

6D/E2

6D/E5

Technology has helped people with disabilities survive and live more conventional lives. 6A/E4** (BSL)

People can learn about others from direct experience, from the mass communications media, and from listening to other people talk about their work and their lives. 7A/E1*

Interaction of Technology and Society (v. 1, p. 37)

Like language, ritual, commerce, and the arts, technology is an intrinsic part of human culture, and it both shapes society and is shaped by it. 3C/E1b

Computers (v. 1, p. 111)

3A/E2

Human beings have made tools and machines to sense and do things that they could not otherwise sense or do at all, or as quickly, or as well. 6A/E2*

People have invented devices such as paper and ink, engraved plastic disks, and magnetic tapes for recording information. These devices enable great amounts of information to be stored, retrieved, and sent to other people or places. 8D/E3

Heredity and Experience Shape Behavior (v. 1, p. 97)
Culture Affects Behavior (v. 1, p. 99)

8F/E2*

5C/E2a

3A/E4

K-2

3A/P1

Technology and Science (p. 17)

3C/P1

People need water, food, air, waste removal, and a particular range of temperatures in their environment, just as other animals do. 6A/P2

5A/P1

People have different external features, such as the size, shape, and color of hair, skin, and eyes, but they are more like one another than like other animals. 6A/P1

People can learn from each other by telling and listening, showing and watching, and imitating what others do. 6D/P3

Communication Technology (v. 1, p. 109)

Cell Functions (v. 1, p. 73)
Use of Earth's Resources (p. 23)
Basic Functions (p. 41)

People have always tried to communicate with one another. Signed and spoken language was one of the first inventions. 8D/P3** (BSL)

Early forms of recording messages used markings on materials such as wood or stone. 8D/P4** (BSL)

similarity to other species

physical attributes of the human species

behavioral attributes of the human species

use of technology

# HUMAN DEVELOPMENT (6B)

Human development, including fertilization, followed by birth, growth, and finally aging and death, continues from one generation to the next. Perhaps no other topic brings individuals closer to a sense of connectedness to people of all cultures and all times. Considerable amounts of society's resources go toward developing technology to control birth and death. The options opened up by technology raise ethical dilemmas for both individuals and society.

The map is organized around four strands—*risks and survival, growth and development, cell differentiation,* and *fertilization and contraception*. The progression of understanding begins in the elementary grades with observable aspects of human development—such as birth and maternal care. The focus in middle school is on reproduction, risks during development, and changes with aging. In high school, the emphasis is on issues raised by reproductive technologies and more abstract ideas about cell differentiation.

In addition to close links to other maps in Chapter 6: THE HUMAN ORGANISM, this map also draws on and contributes to maps in Chapter 5: THE LIVING ENVIRONMENT.

## NOTES

An important thread in the *risks and survival* strand is that, beginning with fertilization, all of human development is affected by factors such as sanitation, diet, and genes. The K-2 benchmark "A human baby..." and the 3-5 benchmark "It takes about nine months..." in the *growth and development* strand deal with the importance of prenatal and infant care and hence contribute also to the *risks and survival* strand.

Two 9-12 benchmarks, "Following fertilization..." and "As successive generations..." in the *cell differentiation* strand, also appear in the *Atlas 1* map **CELLS AND ORGANS**, where the focus is on differentiation and special functions of cells, and in another *Atlas 1* map **DNA AND INHERITED CHARACTERISTICS**. In the *Atlas 1* maps, the benchmark "Following fertilization..." appears in middle-school. It has been moved on this map so that fertilization and cell differentiation are learned together.

Two other 9-12 benchmarks "The availability of artificial means..." and "The development and use..." focus on the social issues raised by reproductive and life-prolonging technologies. Additional benchmarks on the use and impact of health-related technologies, including gene technology, appear in the **HEALTH TECHNOLOGY** map.

### RESEARCH IN *BENCHMARKS*

By the end of 5th grade, students know that babies result from the fusion of sperm and eggs. However, they often don't understand how the fusion brings new life. Before students have an early understanding of genetics, they may believe that the baby exists in the sperm but requires the egg for food and protection, or that the baby exists in the egg and requires the sperm as trigger to growth (Bernstein & Cowan, 1975; Goldman & Goldman, 1982).

Lower elementary-school children understand that death is irreversible and inevitable (Lazar & Torney-Purta, 1991). They usually think death is caused by an external agent but do not connect death with what happens within the body as a result of these external events (Carey, 1985). Around 3rd or 4th grade, students understand that death means the cessation of bodily functions (Carey, 1985).

## 9-12

7F/H2a* 7F/H1 3C/H3* 3C/H1 8F/H7

The development and use of technologies to sustain, prolong, or terminate life raise social, moral, ethical, and legal issues. 6B/H4*

3C/M6* 7E/M3* 7D/H1

As successive generations of an embryo's cells form by division, small differences in their immediate environments cause them to develop slightly differently, by activating or inactivating different parts of the DNA information. 6B/H1

Biological Evolution (v. 1, p. 81)

DNA and Inherited Characteristics (v. 1, p. 69)
Cells and Organs (v. 1, p. 75)

Patterns of human development are similar to those of other vertebrates. 6B/H7** (BSL)

7F/H2a* 7F/H1* 3C/H3* 3C/H1

The availability of artificial means to prevent or facilitate pregnancy raises social, moral, ethical, and legal issues. 6B/H2*

5A/E1

Both genes and environmental factors influence the rate and extent of development. 6B/H5** (SFAA)

Basic Functions (p. 41)

Following fertilization, cell division produces a small cluster of cells that embeds itself in the wall of the uterus. As the embryo develops, it receives nourishment and eliminates wastes by the transfer of substances between its blood and the blood of its mother. 6B/H6** (BSL)

7A/H1

5C/H6 6F/H6** 6E/H2

## 6-8

7D/M1 3C/M4* 8F/M1*

The length and quality of human life are influenced by genes and environmental factors, including sanitation, diet, medical care, and personal health behaviors. 6B/M5d*

Development occurs with somewhat different timing for different individuals. 6B/M7** (SFAA)

7E/M3*

Contraception measures may incapacitate sperm, block their way to the egg, prevent the release of eggs, or prevent the fertilized egg from implanting successfully. 6B/M2

Various body changes occur as adults age. Muscles and joints become less flexible, bones and muscles lose mass, energy levels diminish, and the senses become less acute. Women stop releasing eggs and hence can no longer reproduce. 6B/M5abc

Development sometimes involves dramatic biological changes. For example, puberty involves the maturation of the body to enable reproduction. 6B/M6** (SFAA)

5A/M4*

The developing embryo—and later the newborn infant—is subject to many risks from infection, faults in its genes, its mother's inadequate diet, or her use of cigarettes, alcohol, or other drugs. Inadequate child care may lead to lower physical and mental ability. 6B/M4*

Disease (v. 1, p. 87)
Maintaining Good Health (v. 1, p. 89)

Human fertilization occurs when sperm cells from a male's testes are deposited near an egg cell from the female ovary, and one of the sperm cells enters the egg cell. 6B/M1a*

6E/M2

In sexual reproduction, a single specialized cell from a female merges with a specialized cell from a male. 5B/M2a

5C/M1a*

Basic Functions (p. 41)

## 3-5

DNA and Inherited Characteristics (v. 1, p. 69)
Variation in Inherited Characteristics (v. 1, p. 71)

Human beings live longer than most other animals, but all living things die. 6B/E2

5E/E3

It takes about nine months for a human to develop inside its mother. The developing baby is nourished by the mother, so substances she takes in will affect how well or poorly the baby develops. 6B/E1*

In the first few years after birth, most children make remarkable gains in their physical and mental abilities that allow them to interact with others and with their environment. 6B/E3*

## K-2

A human baby grows inside its mother until its birth. Even after birth, a human baby is unable to care for itself, and its survival depends on the care it receives from adults. 6B/P2

DNA and Inherited Characteristics (v. 1, p. 69)
Variation in Inherited Characteristics (v. 1, p. 71)
Natural Selection (v. 1, p. 83)

All kinds of animals have offspring, usually with two parents involved. 6B/P1*

risks and survival | growth and development | cell differentiation | fertilization and contraception

# BASIC FUNCTIONS (6C)

Like other organisms, human beings are composed of specialized cells grouped in organs that have special functions. The focus of this map is on the essential requirements for life—obtaining food and deriving energy from it, protecting against injury, providing internal coordination, and reproduction—rather than on the body systems as structural entities. Moreover, human body systems and their subsystems illustrate important aspects of systems in general.

The map is organized around four strands, each emphasizing a different survival requirement—*deriving energy from food, defense, coordination,* and *reproduction*. In the elementary grades, the focus is on the needs of the human organism and on body parts that help the body meet its needs. In middle school, the focus is on the role of body systems, cells, and a few simple molecules in carrying out functions to serve survival needs of the organism. High-school benchmarks supply additional detail about the molecules involved in these functions and their role in integrating the work of the various body systems.

At each level of biological organization, a number of connections can be made to maps in *Atlas 1*: to **MAINTAINING GOOD HEALTH** for benchmarks at the elementary- and middle-school levels, to **CELL FUNCTIONS** and **ATOMS AND MOLECULES** for benchmarks at the middle-school level, and to **SYSTEMS** for benchmarks at the high-school level. The *deriving energy from food* strand is closely related to the *Atlas 1* maps on **FLOW OF MATTER IN ECOSYSTEMS** and **FLOW OF ENERGY IN ECOSYSTEMS.** Likewise, the *reproduction* strand has links to the **VARIATION IN INHERITED CHARACTERISTICS** map (also in *Atlas 1*).

## NOTES

The 6-8 benchmark "To burn food for the release of energy…" in the *deriving energy from food* strand describes contributions of the digestive, respiratory, and circulatory systems to this critical basic function. It draws on ideas about energy transformations, burning as a chemical reaction, organs providing cells with basic needs, and interactions and feedback in systems.

Many ideas about basic functions are synthesized in the 6-8 benchmark "Like other animals, human beings…." Students' understanding of body systems at this level depends on their having familiarity with ideas about systems thinking. In turn, benchmarks about body systems and their subsystems illustrate important aspects of systems in general and contribute to a more sophisticated understanding of systems at the 9-12 level.

Starting at middle school, students can relate knowledge about organs and organ systems to their growing understanding of cells. The organs serve the needs of cells and the specialization of cells serves the operation of organs. The new 9-12 benchmark "The human body is a complex system…" builds on the preceding benchmark "Like other animals, human beings…" and introduces more sophisticated understandings about body systems and links ideas about special functions of different kinds of cells to basic functions of the human organism.

## RESEARCH IN *BENCHMARKS*

Lower elementary-school students may have little knowledge about internal bodily organs and think the contents of the body are what they have seen being put into or coming out of it (e.g., food, blood). Upper elementary students can list a large number of organs (Gellert, 1962); however, a sizeable proportion of adults has little knowledge of internal organs or their location (for example, few adults can draw the stomach and the liver in reasonable positions) (Blum, 1977).

By the end of 2nd grade, students know that thought is needed for different kinds of activities (e.g., motor acts) and as a result, know the brain is required for these activities (Carey, 1985). Fourth-graders know the brain helps the body parts but do not always realize that the body also helps the brain (Johnson & Wellman, 1982). Whether upper elementary-school students can achieve this understanding with adequate instruction needs further investigation. Upper elementary-school students attribute to nerves the functions of conducting messages, controlling activity, and stabilizing the body (Gellert, 1962), but even after traditional instruction about the brain and the nervous system, 5th-grade students appear not to understand yet the role of the brain in controlling involuntary behavior (Johnson & Wellman, 1982).

Lower elementary-school students know about circulation and something of the blood's relation to breathing. Upper elementary-school students realize that the heart is a pump, but they are not aware that the blood returns to the heart (Carey, 1985). Students of all ages hold wrong ideas about the structure and function of blood, the structure and function of the heart, the circulatory pattern, the circulatory/respiratory relationships, and the closed system of circulation. Misconceptions concerning the circulatory pattern, the circulatory/respiratory relationships, and the closed system of circulation are difficult to change (Arnaudin & Mintzes, 1985, 1986).

Lower elementary-school students know food is related to growing and being strong and healthy, but they are not aware of the physiological mechanisms. By 5th grade, students know that food undergoes a process of transformation in the body (Contento, 1981; Wellman & Johnson, 1982).

Lower elementary-school students may not know what happens to air after it is inhaled. Upper elementary-school students associate the lungs' activities with breathing and may understand something about the exchange of gases in the lungs and that the air goes to all parts of the body (Carey, 1985).

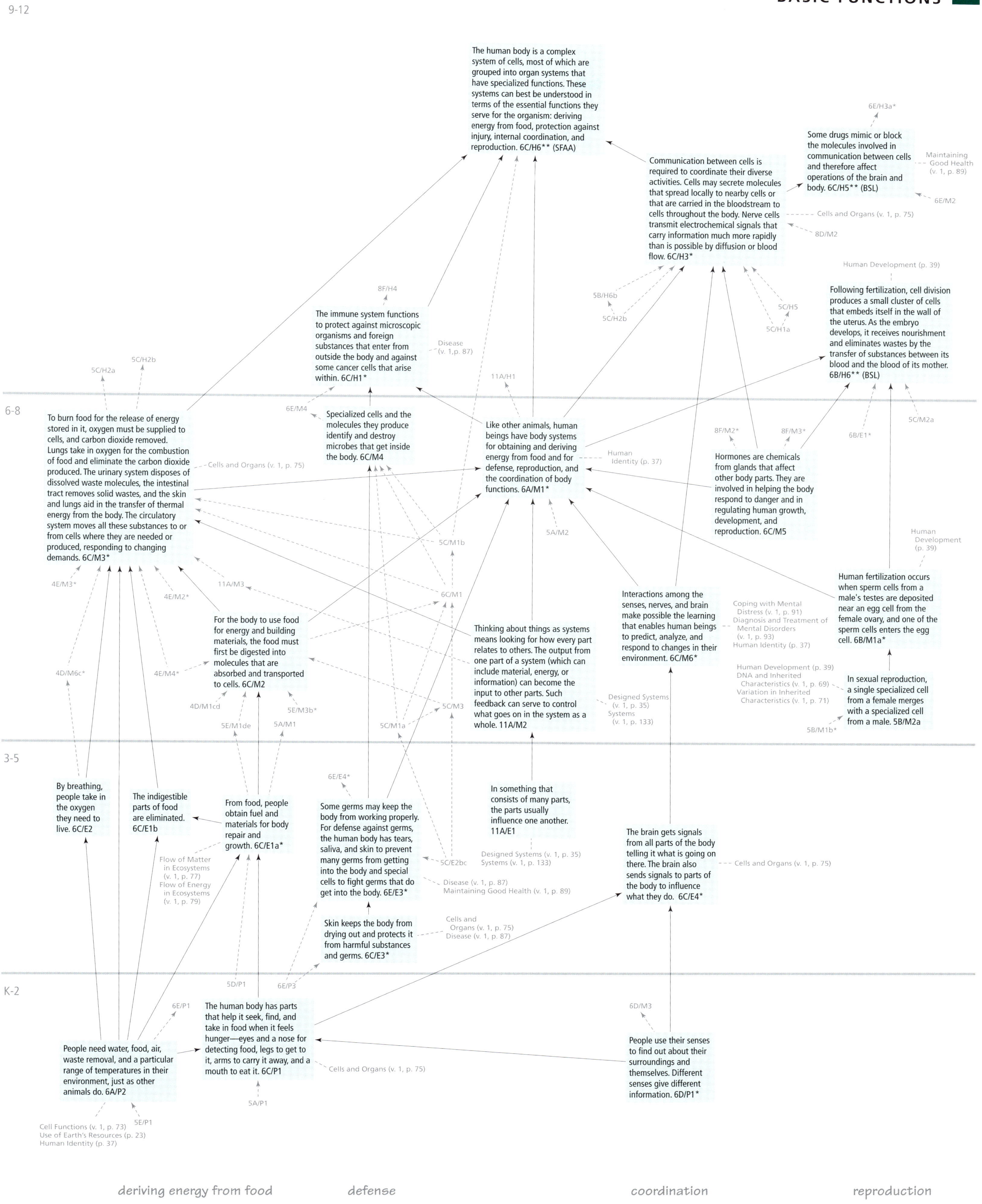
9-12
The human body is a complex system of cells, most of which are grouped into organ systems that have specialized functions. These systems can best be understood in terms of the essential functions they serve for the organism: deriving energy from food, protection against injury, internal coordination, and reproduction. 6C/H6** (SFAA)
6E/H3a*
Some drugs mimic or block the molecules involved in communication between cells and therefore affect operations of the brain and body. 6C/H5** (BSL)
Maintaining Good Health (v. 1, p. 89)
6E/M2
Communication between cells is required to coordinate their diverse activities. Cells may secrete molecules that spread locally to nearby cells or that are carried in the bloodstream to cells throughout the body. Nerve cells transmit electrochemical signals that carry information much more rapidly than is possible by diffusion or blood flow. 6C/H3*
Cells and Organs (v. 1, p. 75)
8D/M2
Human Development (p. 39)
Following fertilization, cell division produces a small cluster of cells that embeds itself in the wall of the uterus. As the embryo develops, it receives nourishment and eliminates wastes by the transfer of substances between its blood and the blood of its mother. 6B/H6** (BSL)
8F/H4
The immune system functions to protect against microscopic organisms and foreign substances that enter from outside the body and against some cancer cells that arise within. 6C/H1*
Disease (v. 1, p. 87)
5B/H6b
5C/H2b
5C/H5
5C/H1a
5C/H2a
5C/H2b
11A/H1
6-8
To burn food for the release of energy stored in it, oxygen must be supplied to cells, and carbon dioxide removed. Lungs take in oxygen for the combustion of food and eliminate the carbon dioxide produced. The urinary system disposes of dissolved waste molecules, the intestinal tract removes solid wastes, and the skin and lungs aid in the transfer of thermal energy from the body. The circulatory system moves all these substances to or from cells where they are needed or produced, responding to changing demands. 6C/M3*
Cells and Organs (v. 1, p. 75)
6E/M4
Specialized cells and the molecules they produce identify and destroy microbes that get inside the body. 6C/M4
Like other animals, human beings have body systems for obtaining and deriving energy from food and for defense, reproduction, and the coordination of body functions. 6A/M1*
Human Identity (p. 37)
8F/M2*
8F/M3*
Hormones are chemicals from glands that affect other body parts. They are involved in helping the body respond to danger and in regulating human growth, development, and reproduction. 6C/M5
6B/E1*
5C/M2a
5C/M1b
5A/M2
Human Development (p. 39)
4E/M3*
4E/M2*
11A/M3
6C/M1
Human fertilization occurs when sperm cells from a male's testes are deposited near an egg cell from the female ovary, and one of the sperm cells enters the egg cell. 6B/M1a*
For the body to use food for energy and building materials, the food must first be digested into molecules that are absorbed and transported to cells. 6C/M2
Thinking about things as systems means looking for how every part relates to others. The output from one part of a system (which can include material, energy, or information) can become the input to other parts. Such feedback can serve to control what goes on in the system as a whole. 11A/M2
Interactions among the senses, nerves, and brain make possible the learning that enables human beings to predict, analyze, and respond to changes in their environment. 6C/M6*
Coping with Mental Distress (v. 1, p. 91)
Diagnosis and Treatment of Mental Disorders (v. 1, p. 93)
Human Identity (p. 37)
4D/M6c*
4E/M4*
Human Development (p. 39)
DNA and Inherited Characteristics (v. 1, p. 69)
Variation in Inherited Characteristics (v. 1, p. 71)
In sexual reproduction, a single specialized cell from a female merges with a specialized cell from a male. 5B/M2a
4D/M1cd
5E/M3b*
5C/M3
Designed Systems (v. 1, p. 35)
Systems (v. 1, p. 133)
5E/M1de
5A/M1
5C/M1a
5B/M1b*
3-5
6E/E4*
By breathing, people take in the oxygen they need to live. 6C/E2
The indigestible parts of food are eliminated. 6C/E1b
From food, people obtain fuel and materials for body repair and growth. 6C/E1a*
Some germs may keep the body from working properly. For defense against germs, the human body has tears, saliva, and skin to prevent many germs from getting into the body and special cells to fight germs that do get into the body. 6E/E3*
In something that consists of many parts, the parts usually influence one another. 11A/E1
The brain gets signals from all parts of the body telling it what is going on there. The brain also sends signals to parts of the body to influence what they do. 6C/E4*
Flow of Matter in Ecosystems (v. 1, p. 77)
Flow of Energy in Ecosystems (v. 1, p. 79)
5C/E2bc
Designed Systems (v. 1, p. 35)
Systems (v. 1, p. 133)
Cells and Organs (v. 1, p. 75)
Disease (v. 1, p. 87)
Maintaining Good Health (v. 1, p. 89)
Skin keeps the body from drying out and protects it from harmful substances and germs. 6C/E3*
Cells and Organs (v. 1, p. 75)
Disease (v. 1, p. 87)
K-2
5D/P1
6E/P3
6E/P1
6D/M3
The human body has parts that help it seek, find, and take in food when it feels hunger—eyes and a nose for detecting food, legs to get to it, arms to carry it away, and a mouth to eat it. 6C/P1
People need water, food, air, waste removal, and a particular range of temperatures in their environment, just as other animals do. 6A/P2
People use their senses to find out about their surroundings and themselves. Different senses give different information. 6D/P1*
Cells and Organs (v. 1, p. 75)
5A/P1
5E/P1
Cell Functions (v. 1, p. 73)
Use of Earth's Resources (p. 23)
Human Identity (p. 37)
deriving energy from food
defense
coordination
reproduction

George Catlin *A Little Sioux Village,* 1861/1869

*As a species, we are social beings who live out our lives in the company of other humans. We organize ourselves into various kinds of social groupings, such as nomadic bands, villages, cities, and countries, in which we work, trade, play, reproduce, and interact in many other ways. Unlike other species, we combine socialization with deliberate changes in social behavior and organization over time. Consequently, the patterns of human society differ from place to place and era to era and across cultures, making the social world a very complex and dynamic environment.*

*Insight into human behavior comes from many sources. The views presented here are based principally on scientific investigation, but it should also be recognized that literature, drama, history, philosophy, and other nonscientific disciplines contribute significantly to our understanding of ourselves. Social scientists study human behavior from a variety of cultural, political, economic, and psychological perspectives, using both qualitative and quantitative approaches. They look for consistent patterns of individual and social behavior and for scientific explanations of those patterns. In some cases, such patterns may seem obvious once they are pointed out, although they may not have been part of how most people consciously thought about the world. In other cases, the patterns—as revealed by scientific investigation—may show people that their long-held beliefs about certain aspects of human behavior are incorrect.*

SCIENCE FOR ALL AMERICANS

# CHAPTER 7

# HUMAN SOCIETY



Social science is a collection of disciplines, each of which examines human behavior from a different perspective and has its own particular techniques, modes of expression, and history. The social-science disciplines employ the evidence-based, hypothesis-testing, model-building approach of science in general to investigate social phenomena, using both qualitative and quantitative approaches, and they share the values characteristic of all science. This chapter describes how various fields of social science illuminate the cultural, political, economic, and psychological dimensions of human behavior.

The list on the left of this page shows the maps for this chapter and indicates the volume of *Atlas of Science Literacy* in which each appears. Those in green typeface are in *Volume 2* and those in gray are in *Volume 1*.

# GROUP BEHAVIOR (7B)

Affiliation with any social group, whether one joins it voluntarily or is born into it, often affects how people inside the group behave toward others and how people outside the group behave toward those inside. Group membership often has advantages for group members, but it can also create a sense of separation between them and those outside the group. Affiliation with a group sometimes leads group members to do and say things, both good and bad, that they would not do or say on their own.

The map has two strands—*group identity* and *group action*. Most of the ideas on this map are introduced in the elementary and middle grades, where many aspects of group membership are understandable to students from personal experience. In the elementary grades, the focus is on why people join groups, the fact that groups have expectations for how their members behave, and that members sometimes behave differently when they are acting as part of a group than they would act by themselves. In the middle grades the focus is extended to the consequences of group membership in the context of the larger society, introducing ideas about the stereotyping of group and non-group members, the exclusion of outsiders, and the exaggeration of differences between group and non-group members that can lead to conflict between them.

**NOTES**

Several benchmarks on this map have been revised to further clarify their meaning and present a more coherent progression. For example, the original wording of benchmark 7B/H2 noted that groups might have unstated purposes, "such as excluding certain categories of people from activities." This example of an unstated activity might have led people to infer that all unstated purposes are reprehensible. The new wording of the benchmark makes it clear that groups can have both official and unofficial purposes, while the issue of groups being exclusionary is covered in benchmark 7B/M1 and elsewhere.

Two new benchmarks have been developed for the *group identity* strand. The benchmark "People join groups..." introduces elementary-level students to reasons for group membership while the middle-school benchmark "Some groups result..." notes that groups differ widely in their structure and nature.

In the *group action* strand the new 9-12 benchmark 7B/H3 is based on text from *Science for All Americans* and addresses the issue of assigning responsibility for group actions.

**RESEARCH IN *BENCHMARKS***

As children try to understand biological and social phenomena, they often overgeneralize information about racial and cultural differences. One must be cautious, however, not to assume that children are prejudiced or deliberately using stereotypes when they overgeneralize. They may simply be thinking typically for young children trying to make sense out of their limited experience with other groups (Ramsey, 1986). Research indicates that stereotypic attitudes begin to develop about 7th grade (Stone, 1986).

Research supports the view that lower elementary-school children are aware of some of the many ways in which rules vary. For example, children agree that some rules in their culture are more important than others and that some rules are more universally right than others. In addition, children are aware of the social function of different kinds of rules (Edwards, 1986). They may go through alternating developmental periods of "affirming" versus "negating" social conventions (Turiel, 1983). Only near the end of middle school and the beginning of high school do students start to accept the need for social conventions to facilitate social interactions in their groups (Mackey, 1991).

*"I'm a social scientist, Michael. That means I can't explain electricity or anything like that, but if you ever want to know about people I'm your man."*

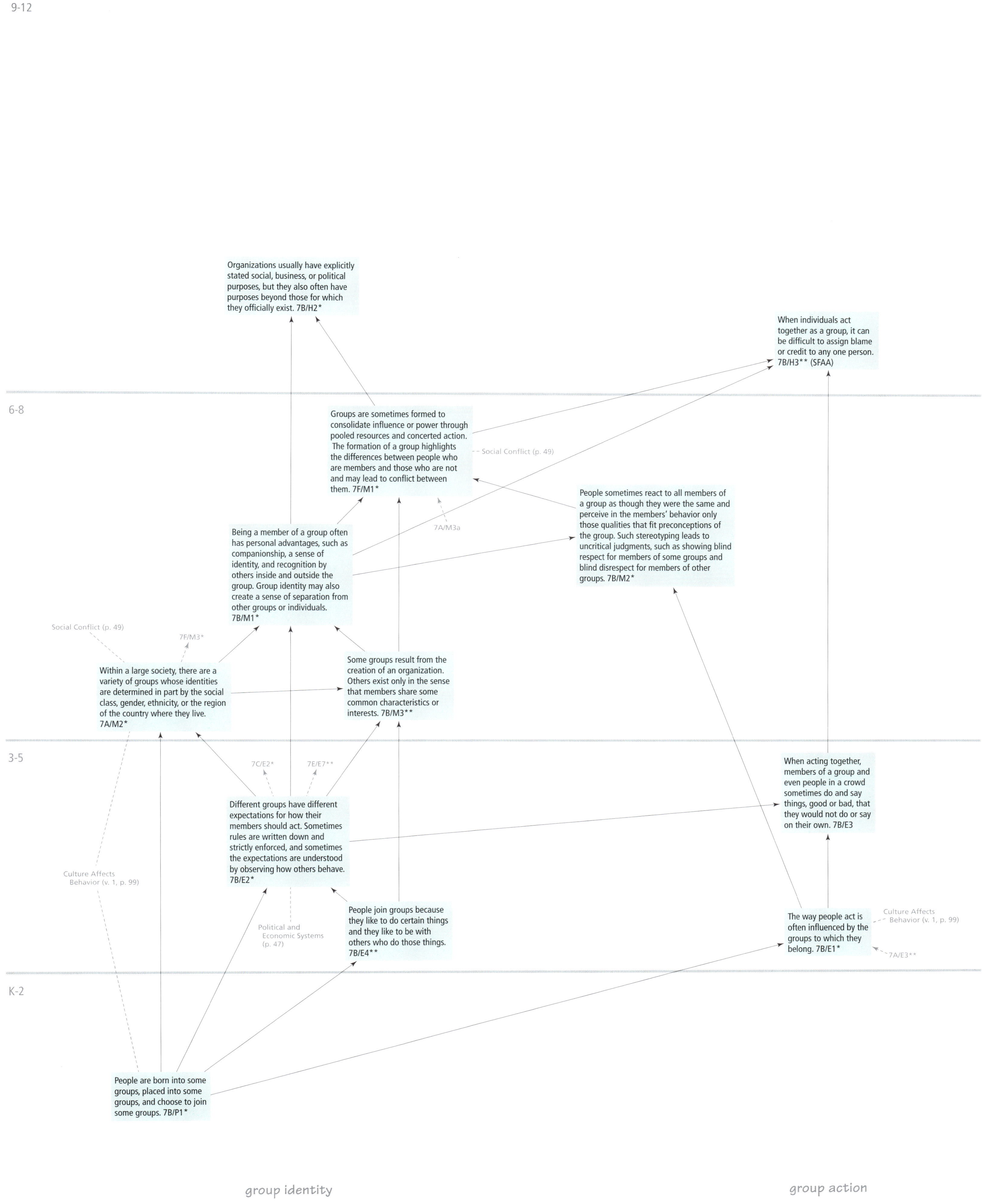
9-12
Organizations usually have explicitly stated social, business, or political purposes, but they also often have purposes beyond those for which they officially exist. 7B/H2*
When individuals act together as a group, it can be difficult to assign blame or credit to any one person. 7B/H3** (SFAA)
6-8
Groups are sometimes formed to consolidate influence or power through pooled resources and concerted action. The formation of a group highlights the differences between people who are members and those who are not and may lead to conflict between them. 7F/M1*
Social Conflict (p. 49)
7A/M3a
People sometimes react to all members of a group as though they were the same and perceive in the members' behavior only those qualities that fit preconceptions of the group. Such stereotyping leads to uncritical judgments, such as showing blind respect for members of some groups and blind disrespect for members of other groups. 7B/M2*
Being a member of a group often has personal advantages, such as companionship, a sense of identity, and recognition by others inside and outside the group. Group identity may also create a sense of separation from other groups or individuals. 7B/M1*
Social Conflict (p. 49)
7F/M3*
Within a large society, there are a variety of groups whose identities are determined in part by the social class, gender, ethnicity, or the region of the country where they live. 7A/M2*
Some groups result from the creation of an organization. Others exist only in the sense that members share some common characteristics or interests. 7B/M3**
3-5
7C/E2*
7E/E7**
When acting together, members of a group and even people in a crowd sometimes do and say things, good or bad, that they would not do or say on their own. 7B/E3
Different groups have different expectations for how their members should act. Sometimes rules are written down and strictly enforced, and sometimes the expectations are understood by observing how others behave. 7B/E2*
Culture Affects Behavior (v. 1, p. 99)
Political and Economic Systems (p. 47)
People join groups because they like to do certain things and they like to be with others who do those things. 7B/E4**
The way people act is often influenced by the groups to which they belong. 7B/E1*
Culture Affects Behavior (v. 1, p. 99)
7A/E3**
K-2
People are born into some groups, placed into some groups, and choose to join some groups. 7B/P1*
group identity
group action

# POLITICAL AND ECONOMIC SYSTEMS (7E)

Societies have systems for allocating power and authority to various groups and individuals who then make, interpret, and enforce the rules and decisions that determine how the society is run. Through their governments, societies make and enforce rules that encompass a wide range of human affairs, one of which is the society's economic activity. For this reason, a nation's political system is often very closely intertwined with its economic system. Although economic systems can vary from nation to nation, almost all of them have some industries that are privately owned and some that are run by the government, and even when privately owned, they usually experience some degree of government regulation.

The map has two strands—*politics* and *economics*. All of the ideas for the *politics* strand are introduced in the elementary and middle grades. In the elementary grades, the focus is on the need for rules and how rules are established and enforced—by parents and teachers as well as by elected officials—and in the middle grades the focus is on the role that formal governments play in the regulation of societies. For the *economics* strand, the focus in the elementary grades is on how the value of things is established and on the use of money as a convenient way of trading for things people want. At the middle-school level, ideas about public and private ownership of industries are explored, and in high school those ideas are extended to include the advantages and disadvantages of government-run and free-market economic systems.

### NOTES

The *politics* strand introduces three important middle-school ideas about the functions of government and the ways in which government leadership is established. These ideas are supported at the elementary level by ideas—revised, in some cases, to improve their clarity—drawn from several sections of *Benchmarks* Chapter 7: HUMAN SOCIETY. In addition, a new benchmark "In some groups, decisions are made by and disputes settled by recognized authorities…" has been developed as a precursor to benchmark 7E/M2.

At the elementary level, the *economics* strand includes several new benchmarks that have been developed to establish the idea of trading for goods and services. Later in the strand, benchmarks focus on the ways in which a system of trading can be organized. Substantial revisions to these benchmarks have been made, particularly at the high-school level. These revisions emphasize that most nations opt for a compromise between total government control and total individual freedom in economic matters.

Finally, the level of interaction between political and economic systems in a nation is described in benchmark 7E/M6, a new statement based on text in *Science for All Americans* that contributes to both the *politics* and *economics* strands.

### RESEARCH IN *BENCHMARKS*

Much of the research related to this topic relies on samples of students outside the United States. These studies need to be replicated with U.S. students from different social backgrounds because research suggests that students' environments may influence their economic ideas. Lower elementary-school students have already formed a fairly accurate concept of scarcity. However, the majority of children at that age may still reason in a superficial and literal manner about economic problems concerning choice and trade-offs (Schug & Birkey, 1985).

Elementary-school students know that workers are paid by bosses for the work done. They also know that consumers pay shopkeepers and that shopkeepers pay middlemen and producers in exchange for goods. However, until 4th or 5th grade they do not connect these two areas of experience. As a result, they may believe that the boss must have access to sources external to the factory or store to pay workers. Or they may think that prices remain the same or even decrease in the passage from producer to consumer (Berti & Bombi, 1988).

The concept of profit, which is a prerequisite idea to understanding the notion of a market economy, emerges naturally in children between the upper elementary and the early middle-school grades (Jahoda, 1979, 1981). By 4th grade, students can usually understand the notion of profit, after specially designed instruction. Students' ability to compare expenses (total costs) and revenue is a prerequisite to this understanding (Berti, 1992).

There are several student difficulties related to understanding the price mechanism in market economies. Even at the college level, students have difficulty understanding that price is not an inherent characteristic of goods but a function of demand and supply (Berti & Grivet, 1990; Marton,1978). Students often fail to connect the different viewpoints of consumers and producers. For example, while believing that consumers buy less when prices increase, some students may also believe the reverse and that, if demand drops, producers can increase prices to earn the same amount of money as before (Berti & Grivet, 1990).

Lower elementary-school students do not have a notion of production based on the transformation of primary materials, for example, furniture from wood. In addition, students at this age have little knowledge about agricultural and industrial production. They believe that farmers themselves consume all the products from their crops and animals. It is only in 2nd grade that the majority of the children recognize the existence of a producer distinct from a shopkeeper (Berti & Bombi, 1988).

Elementary-school children typically do not understand public institutions as institutions that provide collective services. For them, terms like "council," "state," or "government" do not specifically designate any particular body, nor do they use the terms in a sense that distinguishes them clearly from a private employer. Instead, these terms are perceived as referring to important or wealthy persons who wield authority and pay people who work (Berti & Bombi, 1988). Whether young children can understand these concepts with adequate instruction needs further investigation. Research indicates students acquire the intellectual capacity to construct a political order in hierarchical form at around 5th grade (Connell, 1971). From this age on, various authorities are no longer thought to have power over only a few persons close to them, but over whole populations through the enactment of laws and the control of power. By this age, students also know that political parties exist and their activities affect elections (Mackey, 1991).

## 9-12

Governments can use a variety of strategies to keep a free-market system functioning efficiently and to protect individuals and the society as a whole from the negative effects of free-market behavior. Interventions that protect individuals include placing restrictions on unfair hiring, requiring private industries to offer a minimum wage and make health insurance available to their workers, and providing various forms of welfare support to those who cannot pay for basic services themselves. Interventions that protect society at large include imposing tax rates that increase with wealth and placing restrictions on environmental pollution, unfair advertising, and the sale of unsafe products. Governments maintain competition among industries in a free-market system by restricting the establishment of monopolies and by providing support to certain critical industries to maintain their ability to compete. 7E/H3*

7G/H4*

In almost all societies, governments run some industries for the benefit of the society as a whole. The advantage of this approach compared to the free-market approach is that goods and services that the society wants can be made available to citizens even if it is not profitable for private industry to offer them. The disadvantage is that decisions must be made by government officials, often with input from the public, on the basis of factors that are much more complex than profit alone. In addition, there may be less incentive to work as efficiently when the profit motive is absent and the system does not respond as quickly to short-term variations in supply and demand that the free-market system allows. 7E/H2*

In a free-market economic system, it is assumed that free competition produces the best allocation of resources, the greatest productivity, and the lowest costs. Decisions are made naturally as consumers and businesses interact in the marketplace. Free-market enterprises are initiated by individuals or groups of individuals who invest and risk their own resources, or resources they borrow, in order to earn profits on that invested capital. The anticipation of profit motivates individual initiative, innovation, and hard work in a free-market economy. Problems may arise if one business achieves a monopoly in the delivery of a particular product or service, so that the motivation to deliver that product or service efficiently is no longer present. And if all decisions are determined by an unregulated market system, some individuals may become extremely rich and others so poor that they can barely afford even the basic necessities such as food and shelter. 7E/H1*

7C/H4

## 6-8

Governments are run by people who come into power by election, appointment, or force. 7E/M2*

Influences on Social Change (v. 1, p. 101)
Social Decisions (v. 1, p. 103)

7F/E4

7F/E2

7D/M2

The political system of a nation is closely intertwined with its economic system, refereeing the economic activity of individuals and groups at every level. 7E/M6** (SFAA)

Almost all societies have some industries that are privately owned and some that are run by government agencies. 7E/M4c*

Governments usually have most of the power to make, revise, interpret, and enforce the rules and decisions that determine how a community, state, or nation is run. Many of the rules established by governments are designed to reduce social conflict. The rules affect a wide range of human affairs, from marriage and education to scientific research and commerce. 7E/M3*

Some industries are controlled by government agencies that decide which goods or services to provide and how to distribute them based on various political considerations. 7E/M4a*

Some industries are privately owned and controlled by many different individuals or groups of individuals that decide what goods or services to provide based on what they believe will personally benefit them the most. 7E/M4b*

Government provides some goods and services through its own agencies and some through contracts with private individuals or businesses. To pay for the goods and services, government must obtain money by taxing people or by borrowing the money. 7E/M1

Most groups have formal or informal procedures for arbitrating conflicts between their members. 7F/M2*

Social Conflict (p. 49)

The goods and services that people want are supplied by individuals or by groups of individuals. 7E/M5**

## 3-5

8B/E4*

Rules can sometimes be changed by getting whoever made the rules to agree to change them. 7C/E2*

Influences on Social Change (v. 1, p. 101)

In some groups, decisions are made by and disputes settled by recognized authorities, such as parents, teachers, bosses, or elected officials. 7E/E7**

Sometimes a community decides to make certain services—such as schools, libraries, parks, mail service, and police and fire protection— equally available to all its members through government agencies. 7E/E2*

People who live together in a group often reach an agreement about what each will contribute to the others in the group. 7E/E1*

The value of something depends on how much of it is available and how many people want it. 7E/E6**

Social Conflict (p. 49)

Different groups have different expectations for how their members should act. Sometimes rules are written down and strictly enforced, and sometimes the expectations are understood by observing how others behave. 7B/E2*

Group Behavior (p. 45)

For many things they need, people rely on others who are not part of their family and may not even be part of their local community. 7G/E2** (BSL)

Global Interdependence (p. 51)

Trading things people have or things they want for money is often more convenient than trading those things directly. The same is true for services people provide or services that they want. 7E/E5** (BSL)

## K-2

7B/P1

Rules at home, at school, and in the community let people know how to behave and what to expect from others and so can reduce the number of disputes. 7F/P2*

Social Conflict (p. 49)

Influences on Social Change (v. 1, p. 101)

Rules can help to make sure that everyone is treated fairly. 7E/P2*

Social Decisions (v. 1, p. 103)

People often trade things they have for things that they want. 7E/P3**

Global Interdependence (p. 51)

politics

economics

# SOCIAL CONFLICT (7F)

Conflict between individuals and between groups exists in all societies. Conflict can arise from many sources but primarily because of competition for resources, power, or status. Because social conflicts are so common, all societies have established formal and informal ways of resolving them, at the local and family level and at the national level.

The map has two strands—*reasons for conflict* and *resolving conflict*. In the elementary grades benchmarks identify a variety of positive strategies for resolving conflict—making different points of view clear, compromising, and seeking outside help. In the middle grades, the focus is on how rules established by organizations and by governments are designed to reduce social conflict. In high school, attention shifts to the difficulties involved in resolving conflicts, even when there are established mechanisms in place for doing so. At all grade levels, a variety of reasons are introduced for why conflicts arise.

Because social conflict is sometimes the result of human behavior, benchmarks on this map have strong connections to other maps in Chapter 7, especially maps on **GROUP BEHAVIOR** in this volume and **CULTURE AFFECTS BEHAVIOR** in *Atlas 1*.

## NOTES

There are several new benchmarks in the *reasons for conflict* strand, including the 3-5 benchmark "Different cultures have different patterns of behavior...," which also appears on the **CULTURE AFFECTS BEHAVIOR** map in *Atlas 1*. To set the stage for a more sophisticated consideration of conflict resolution in high school, the benchmark "Conflict between people or groups...," originally a 9-12 benchmark, has been moved to the middle-school level. By making this change, students will have some examples of common reasons for conflict before they try to understand how conflicts might be resolved. Another new 6-8 benchmark "Social change, or the prospect of it..." adds depth to ideas about the role that social changes play in producing conflicts.

In the *resolving conflict* strand, a number of benchmarks have been revised to improve their clarity and increase the coherence of the strand. For example, the original wording of benchmark 7F/E1 indicated that communicating different points of view in a dispute helps people find a compromise. The new version notes that this communication is simply a first step toward a compromise. A new 9-12 benchmark "Conflicts can be resolved..." describes different methods for settling disputes.

## RESEARCH IN *BENCHMARKS*

Studies on young children's recognition of conflict indicate that from age six on, children are able to recognize that a child's desires may conflict with those of his parents or friends (Berti, 1988; Damon, 1977), but it is not clear whether children at this age can also recognize conflict among adults. Upper elementary-school students may not recognize that making laws (to settle conflicts) is the job of the whole community as well as political leaders (Berti, 1988). Middle-school students may not recognize conflict involving social groups (Berti, 1988; Connell, 1971). Middle-school students do not recognize the role of debate, disagreement, and conflict in the operation of the democratic political system (Hess & Torney, 1968).

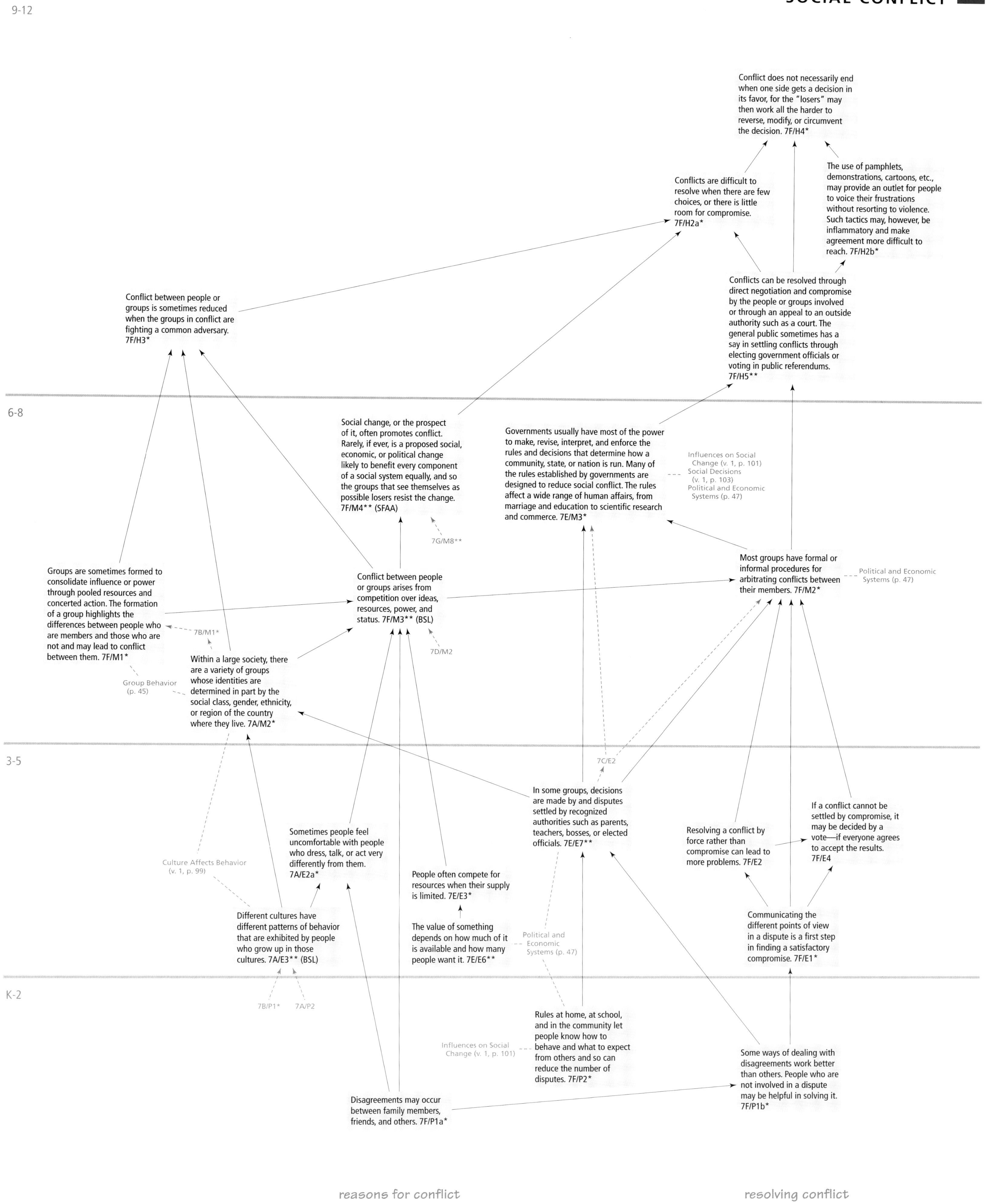
9-12
Conflict does not necessarily end when one side gets a decision in its favor, for the "losers" may then work all the harder to reverse, modify, or circumvent the decision. 7F/H4*
Conflicts are difficult to resolve when there are few choices, or there is little room for compromise. 7F/H2a*
The use of pamphlets, demonstrations, cartoons, etc., may provide an outlet for people to voice their frustrations without resorting to violence. Such tactics may, however, be inflammatory and make agreement more difficult to reach. 7F/H2b*
Conflict between people or groups is sometimes reduced when the groups in conflict are fighting a common adversary. 7F/H3*
Conflicts can be resolved through direct negotiation and compromise by the people or groups involved or through an appeal to an outside authority such as a court. The general public sometimes has a say in settling conflicts through electing government officials or voting in public referendums. 7F/H5**
6-8
Social change, or the prospect of it, often promotes conflict. Rarely, if ever, is a proposed social, economic, or political change likely to benefit every component of a social system equally, and so the groups that see themselves as possible losers resist the change. 7F/M4** (SFAA)
Governments usually have most of the power to make, revise, interpret, and enforce the rules and decisions that determine how a community, state, or nation is run. Many of the rules established by governments are designed to reduce social conflict. The rules affect a wide range of human affairs, from marriage and education to scientific research and commerce. 7E/M3*
Influences on Social Change (v. 1, p. 101)
Social Decisions (v. 1, p. 103)
Political and Economic Systems (p. 47)
7G/M8**
Groups are sometimes formed to consolidate influence or power through pooled resources and concerted action. The formation of a group highlights the differences between people who are members and those who are not and may lead to conflict between them. 7F/M1*
Conflict between people or groups arises from competition over ideas, resources, power, and status. 7F/M3** (BSL)
Most groups have formal or informal procedures for arbitrating conflicts between their members. 7F/M2*
Political and Economic Systems (p. 47)
7B/M1*
7D/M2
Group Behavior (p. 45)
Within a large society, there are a variety of groups whose identities are determined in part by the social class, gender, ethnicity, or region of the country where they live. 7A/M2*
3-5
7C/E2
In some groups, decisions are made by and disputes settled by recognized authorities such as parents, teachers, bosses, or elected officials. 7E/E7**
Sometimes people feel uncomfortable with people who dress, talk, or act very differently from them. 7A/E2a*
Resolving a conflict by force rather than compromise can lead to more problems. 7F/E2
If a conflict cannot be settled by compromise, it may be decided by a vote—if everyone agrees to accept the results. 7F/E4
Culture Affects Behavior (v. 1, p. 99)
People often compete for resources when their supply is limited. 7E/E3*
Different cultures have different patterns of behavior that are exhibited by people who grow up in those cultures. 7A/E3** (BSL)
The value of something depends on how much of it is available and how many people want it. 7E/E6**
Political and Economic Systems (p. 47)
Communicating the different points of view in a dispute is a first step in finding a satisfactory compromise. 7F/E1*
K-2
7B/P1*
7A/P2
Rules at home, at school, and in the community let people know how to behave and what to expect from others and so can reduce the number of disputes. 7F/P2*
Influences on Social Change (v. 1, p. 101)
Some ways of dealing with disagreements work better than others. People who are not involved in a dispute may be helpful in solving it. 7F/P1b*
Disagreements may occur between family members, friends, and others. 7F/P1a*
reasons for conflict
resolving conflict

# GLOBAL INTERDEPENDENCE (7G)

Nations and cultures are increasingly interconnected through international travel, international economic systems, the mass media, and shared environmental problems. The wealth, security, and general welfare of almost all nations are interconnected, and although global interdependence often has positive effects, not everyone benefits equally.

The map has three strands—*communication and transportation technologies, trade,* and *treaties*. In the elementary grades, the focus is on how so many of the things that people use do not come from their families or even from their local communities but from other countries. In the middle grades, the focus is on agreements made between countries to maintain peaceful relationships, to trade with each other, and to manage the global environment. In high school, the focus is on the trade-offs and complications that come with globalization.

## NOTES

In the *communication and transportation technologies* strand, benchmarks 7G/M6 and 7G/H5 have been developed to address the increasing role that such technologies play in global interdependence.

A number of the benchmarks in the *trade* strand have been revised to clarify their meaning and to increase the coherence of the strand. For example, the 6-8 benchmark "Trade between nations..." has been revised to take note of differences in human resources as well as differences in natural resources as important influences on trade between nations.

In the *treaties* strand, the 6-8 benchmark 7G/M7 has been developed to point out the role of international organizations in facilitating relationships among nations. Another new middle-school benchmark introduces the idea that "Treaties do not affect all of the people in a country equally."

## RESEARCH IN *BENCHMARKS*

Some research suggests that middle-school and high-school students have an understanding of the global nature of trade, although they have only limited understanding about reciprocal benefits from trade (Schug & Lephardt, 1992).

9-12

International trade is often complicated by political motivations taking priority over economic ones. 7G/H2*

The wealth of a country depends on the balance between how much its resources and products are sought by other nations and how much of other nations' resources and products it seeks. Even if a country could produce everything it needs for itself, it may still benefit from trade with other countries. 7G/H1*

8C/H4*

The growing worldwide interdependence of social, economic, and ecological systems means that changes in one place in the world may have effects in any other place. 7G/H4*

7E/H3*

8C/H5*

Communication and transportation technologies, coupled with political and economic policies, now allow people to interact with people in different countries almost as easily as they interact with people in their own country. This has allowed for the spread of political, economic, and cultural influences across the planet much more rapidly than had been the case in the past. Like any social change, there are trade-offs in the globalization of the planet, and it benefits some people more than others. 7G/H5**

Migration of workers between nations—temporary and permanent, legal and illegal—plays a major role in the quality and availability of the workforce in many nations. It can bring both economic benefits and political problems. 7G/H3*

10J/H1*

6-8

Treaties do not affect all of the people in a country equally. 7G/M8** (SFAA)

7F/M4**

The economic well-being of a country may be enhanced by improving the skill and motivation of its workforce, by developing equipment or practices that are more efficient and produce higher quality products and services, and by engaging in trade with other nations. 7G/M2*

The global environment is affected by national and international policies and practices relating to energy use, waste disposal, ecological management, manufacturing, and population. 7G/M5*

International organizations such as the United Nations, the World Bank, and the European Union provide a structure for nations to communicate, reach consensus, and negotiate treaties, but these organizations often have only limited authority over their members. 7G/M7** (SFAA)

Communication and transportation technologies influence how much people in different countries may interact, whether it be for political, economic, or social purposes. 7G/M6**

8B/M3*

8B/M2*

4B/M11a*

4C/M10**

4C/M7

Treaties are negotiated between two or more nations to establish or maintain peaceful relationships, to define parameters for trade, or to create political or military alliances. 7G/M3*

Trade between nations occurs when natural resources or the skill to make something are unevenly distributed or when the costs of production are very different in different countries. A nation has a trade opportunity whenever it can create more of a product or service or a better quality product or service, or a less expensive product or service than can another nation. 7G/M1*

8B/M4*

8C/M6*

4B/M10ab*

8D/M2

3-5

Although many things are still made by hand in some parts of the world, almost everything in the most technologically developed countries is now produced using machines that are automated. By using machinery, the time required to make a product and its cost can be greatly reduced. 8B/E4*

Manufacturing (p. 57)
The Industrial Revolution (p. 89)

Trade occurs between individual people, between nations, and between regions in the same nation. 7G/E1b*

Communication technologies make it possible to send and receive information more and more reliably, quickly, and cheaply over long distances. 8D/E4

Communication Technology (v. 1, p. 109)

For many things they need, people rely on others who are not part of their family and may not even be part of their local community. 7G/E2** (BSL)

Many of the things people eat and wear come from other countries. 7G/E1a*

K-2

Political and Economic Systems (p. 47)

People often trade things they have for things that they want. 7E/P3**

communication and transportation technologies

trade

treaties

Grant Wood *Fall Plowing,* 1931

*The world we live in has been shaped in many important ways by human action. We have created technological options to prevent, eliminate, or lessen threats to life and the environment and to fulfill social needs. We have dammed rivers and cleared forests, made new materials and machines, covered vast areas with cities and highways, and decided—sometimes willy-nilly—the fate of many other living things.*

*In a sense, then, many parts of our world are designed—shaped and controlled, largely through the use of technology—in light of what we take our interests to be. We have brought the earth to a point where our future well-being will depend heavily on how we develop and use and restrict technology. In turn, that will depend heavily on how well we understand the workings of technology and the social, cultural, economic, and ecological systems within which we live.*

SCIENCE FOR ALL AMERICANS

# CHAPTER 8



# THE DESIGNED WORLD

The term "technology" increasingly seems to be used to refer only to electronic technology. *Science for All Americans* and *Benchmarks for Science Literacy* focus on a more general conception of technology as ways of doing things and having specific effects on the world. Toothpicks, plows, drugs, and tollbooths are as much technology as televisions and computers.

Maps that treat topics related to technology are presented in five chapters. This chapter lays out a progression of understanding of principles relevant to some of the key technologies of today's world; Chapter 3: THE NATURE OF TECHNOLOGY outlines the development of ideas about engineering as a unique endeavor; Chapter 10: HISTORICAL PERSPECTIVES suggests a sequence of ideas to help students understand the Industrial Revolution; Chapter 11: COMMON THEMES maps out powerful engineering concepts such as systems and models that are useful as analytical tools across disciplines; and Chapter 12: HABITS OF MIND identifies the skills needed for participating in a technological world.

The list on the left of this page shows the maps for this chapter and indicates the volume of *Atlas of Science Literacy* in which each appears. Those in green typeface are in *Volume 2* and those in gray are in *Volume 1*.

# MATERIALS SCIENCE (8B)

Technology is based on the use and application of a great variety of materials whose properties determine their potential uses. In recent decades, materials processing has focused increasingly on the synthesis of materials with entirely new properties. The rapid growth of materials processing has lead to the rapid use of materials, raising concerns about the depletion of resources and the disposal of used materials. Solving these problems will require systematic efforts that include both social and technological innovations and, hence, the need for an informed public.

The map is organized around three strands—*physical properties, synthesis and use of materials,* and *resource depletion and recycling*. In the elementary grades, the focus is on properties, uses, and recycling of objects and materials. In middle school, the focus is on the production of new materials through chemical reactions and issues related to resource depletion and disposal of used materials. In high school, the emphasis shifts to using chemical principles to design new materials and to the need for social and technological solutions to issues of depletion and disposal of materials.

This map shares many ideas with maps in Chapter 4: THE PHYSICAL SETTING, particularly the **ATOMS AND MOLECULES** and **CHEMICAL REACTIONS** maps in *Atlas 1* and **USE OF EARTH'S RESOURCES** and **ELECTRICITY AND MAGNETISM** maps in this volume.

## NOTES

In the *physical properties* strand, the progression of understanding begins with differentiating between objects and the materials that make them up, and between properties of objects and properties of materials. In grades 3-5, those properties are further elaborated. In grades 6-8, the concept of material is extended to the formal concept of substance that is identified by fixed and reproducible properties. Two new 9-12 benchmarks focus on the nanoscale: "Objects made up of a small number of atoms…" distinguishes properties of macroscopic objects from those of structures made up of small numbers of atoms, and "Groups of atoms and molecules…" relates differences in properties to differential effects of scale.

The *synthesis and use of materials* strand illustrates the importance of science to technology and the difficulty of clearly separating the two domains. The 9-12 benchmark "Increased knowledge of the properties…" provides a particularly appropriate context for exploring how scientific knowledge fuels technological advances and how technology creates new scientific knowledge.

## RESEARCH IN *BENCHMARKS*

Benchmarks in THE DESIGNED WORLD chapter are associated with knowledge and skills needed for other literacy goals. For example, they draw upon benchmarks in THE MATHEMATICAL WORLD chapter for knowledge of shapes, estimation, measure, and the ability to use scale, and upon benchmarks in THE PHYSICAL SETTING chapter for knowledge of materials and their properties, forces, and energy. As a result, the literature on student understanding of these topics provides some insight into when and how students may understand concepts of THE DESIGNED WORLD. For example, research on student understanding of materials suggests that the tasks of classifying objects according to what they are made of and of comparing properties of materials can be challenging for early elementary-school children. In addition, elementary-school children may have limited knowledge or hold misconceptions about the origins and transformations of materials (Russell, Longden, & McGuigan, 1991).

## 9-12

3A/H4**

Increased knowledge of the properties of particular molecular structures helps in the design and synthesis of new materials for special purposes. 8B/H4*

Manufacturing (p. 57)

11D/H2*

3A/M3*

The development of new materials and the increased use of existing materials by a growing human population have led to the removal of resources from the environment much more rapidly than they can be replaced by natural processes. Disposal of waste materials has also become a problem. Solving these problems requires systematic efforts involving both social and technological innovations. 8B/H7** (SFAA)

Use of Earth's Resources (p. 23)
Manufacturing (p. 57)

Semiconducting materials differ greatly in how well they conduct electrons, depending on the exact composition of the material. 4G/H4d*

At very low temperatures, some materials become superconductors and offer no resistance to the flow of electrons. 4G/H4c*

Electricity and Magnetism (p. 27)

4G/H4ab*

Groups of atoms and molecules can form structures that can be measured in billionths of a meter. The properties of structures at this scale (known as the nanoscale) and materials composed of such structures, can be very different than the properties at the macroscopic scale because of the increase in the ratio of surface area to volume and changes in the relative strengths of different forces at different scales. Increased knowledge of the properties of materials at the nanoscale provides a basis for the development of new materials and new uses of existing materials. 8B/H6**

Waste management includes considerations of quantity, safety, degradability, and cost. It requires social and technological innovations, because waste-disposal problems are political and economic as well as technical. 8B/H2

Objects made up of a small number of atoms may exhibit different properties than macroscopic objects made up of the same kinds of atoms. 8B/H5**

8C/H4*

## 6-8

Atoms may link together in well-defined molecules, or may be packed together in crystal patterns. Different arrangements of atoms into groups compose all substances and determine the characteristic properties of substances. 4D/M1cd*

Some materials, such as plastics, are synthesized in chemical reactions that link atoms together in long chains. Plastics can be designed to have a variety of properties for a variety of uses. 8B/M6** (SFAA)

3C/M5

7G/M5*

4C/M7

The wasteful or unnecessary use of natural resources can limit their availability for other purposes. Restoring depleted soil, forests, or fishing grounds can be difficult and costly. 4B/M11a*

Efforts to find replacements for existing materials are driven by an interest in finding materials that are cheaper to obtain or produce or that have more desirable properties. 8B/M5** (SFAA)

3B/M2a

Atoms and Molecules (v. 1, p. 55)
Conservation of Matter (v. 1, p. 57)
States of Matter (v. 1, p. 59)
Chemical Reactions (v. 1, p. 61)
Flow of Matter in Ecosystems (v. 1, p. 77)
Flow of Energy in Ecosystems (v. 1, p. 79)
The Chemical Revolution (p. 81)

The benefits of Earth's resources—such as fresh water, air, soil, and trees—can be reduced by deliberately or inadvertently polluting them. The atmosphere, the oceans, and the land have a limited capacity to absorb and recycle waste materials. In addition, some materials take a long time to degrade. Therefore, cleaning up polluted air, water, or soil can be difficult and costly. 4B/M11bc*

Use of Earth's Resources (p. 23)

A substance has characteristic properties such as density, a boiling point, and solubility, all of which are independent of the amount of the substance and can be used to identify it. 4D/M10** (NSES)

3B/M1*

Use of Earth's Resources (p. 23)

Recycling materials and the development of substitutes for those materials can reduce the rate of depletion of resources but may also be costly. Some materials are not easily recycled. 4B/M10c*

8F/H9**

The choice of materials for a job depends on their properties. 8B/M1*

7G/M1*

Materials vary in how they respond to electric currents, magnetic forces, and visible light or other electromagnetic waves. 4D/M9**

Design Constraints (v. 1, p. 33)
Manufacturing (p. 57)

Some material resources are very rare and some exist in great quantities. The ability to obtain and process resources depends on where they are located and the form they are in. As resources are depleted, they may become more difficult to obtain. 4B/M10ab*

Scale (p. 99)

Use of Earth's Resources (p. 23)

4E/M6**

8C/M1*

4D/M6a

3C/E4b

## 3-5

Electricity and Magnetism (p. 27)
Using Tools and Devices (p. 109)

Humans have produced a wide variety of materials, such as steel, plastic, and nylon, that do not appear in nature. 8B/E2*

Technology and Science (p. 17)

3B/E1*

7E/E3*

4C/E2

4D/E4*

All materials have certain physical properties, such as strength, hardness, flexibility, durability, resistance to water and fire, and ease of conducting heat. 4D/E6** (SFAA)

Naturally occurring materials such as wood, clay, cotton, and animal skins may be processed to change their properties. 8B/E1*

Sometimes it is possible to use the materials from discarded products to make new products, but materials differ widely in the ease with which they can be recycled. 8B/E3b

Discarded products contribute to the problem of waste disposal. 8B/E3a

4E/E2b*

Ues of Earth's Resources (p. 23)

## K-2

Design Constraints (v. 1, p. 33)
Atoms and Molecules (v. 1, p. 55)
Conservation of Matter (v. 1, p. 57)
Chemical Reactions (v. 1, p. 61)

Design Constraints (v. 1, p. 33)
Manufacturing (p. 57)

Some kinds of materials are better than others for making any particular thing. Materials that are better in some ways, such as stronger or cheaper, may be worse in other ways, such as heavier or harder to cut. 8B/P1*

Some objects can be used over again. 8B/P4*

Objects can be described in terms of their properties. Some properties, such as hardness and flexibility, depend upon what material the object is made of, and some properties, such as size and shape, do not. 4D/P1*

7D/P1

physical properties | synthesis and use of materials | resource depletion and recycling

# MANUFACTURING (8B)

Modern manufacturing is based largely on the need to produce great amounts of products of uniform quality and much smaller amounts of products of extremely high quality. The growth of technology has been greatly helped by improvement in the fineness and sharpness of cutting tools, the force that can be applied, the temperature at which heat can be concentrated, the swiftness with which operations can occur, and the consistency with which operations can be repeated. The evolution of production, which now involves highly complex tasks and automation, has changed the nature of work and requires increased flexibility and skill of workers in learning a succession of new job roles.

The map is organized around four strands—*materials, manipulation skills, manufacturing,* and *societal implications*. In the elementary grades, the focus is on using materials and tools to make things and understanding some of the benefits of mass production. In middle school, the focus is on the role of technology and automation in manufacturing and some of their benefits. In high school, the emphasis shifts to understanding the contribution of science to manufacturing technology and the social implications of technology for society.

This map has links to skills on the **USING TOOLS AND DEVICES** map and to events described on **THE INDUSTRIAL REVOLUTION** map.

## NOTES

Many of the K-8 benchmarks in the *materials* strand appear in or contribute to benchmarks in the **MATERIALS SCIENCE** map in this volume and in the **DESIGN CONSTRAINTS** map in *Atlas 1*.

The skills benchmarks in the *manipulation skills* strand are connected to knowledge benchmarks about materials and manufacturing, but they are also important learning goals in their own right and are essential to students' science literacy.

Benchmarks in the *manufacturing* strand provide opportunities for connecting ideas about manufacturing with trade issues presented in the **GLOBAL INTERDEPENDENCE** map. The 3-5 benchmark "In the 1700s, most manufacturing was still done in homes..." introduces ideas about the development of manufacturing that also appear within the historical context of **THE INDUSTRIAL REVOLUTION** map.

A new 9-12 benchmark "The development of new materials..." addresses environmental implications of the development of new materials. It also appears in and is supported by benchmarks in the **MATERIALS SCIENCE** map, which also includes benchmarks on how manufactured products affect the environment.

## RESEARCH IN *BENCHMARKS*

No relevant research available in *Benchmarks*.

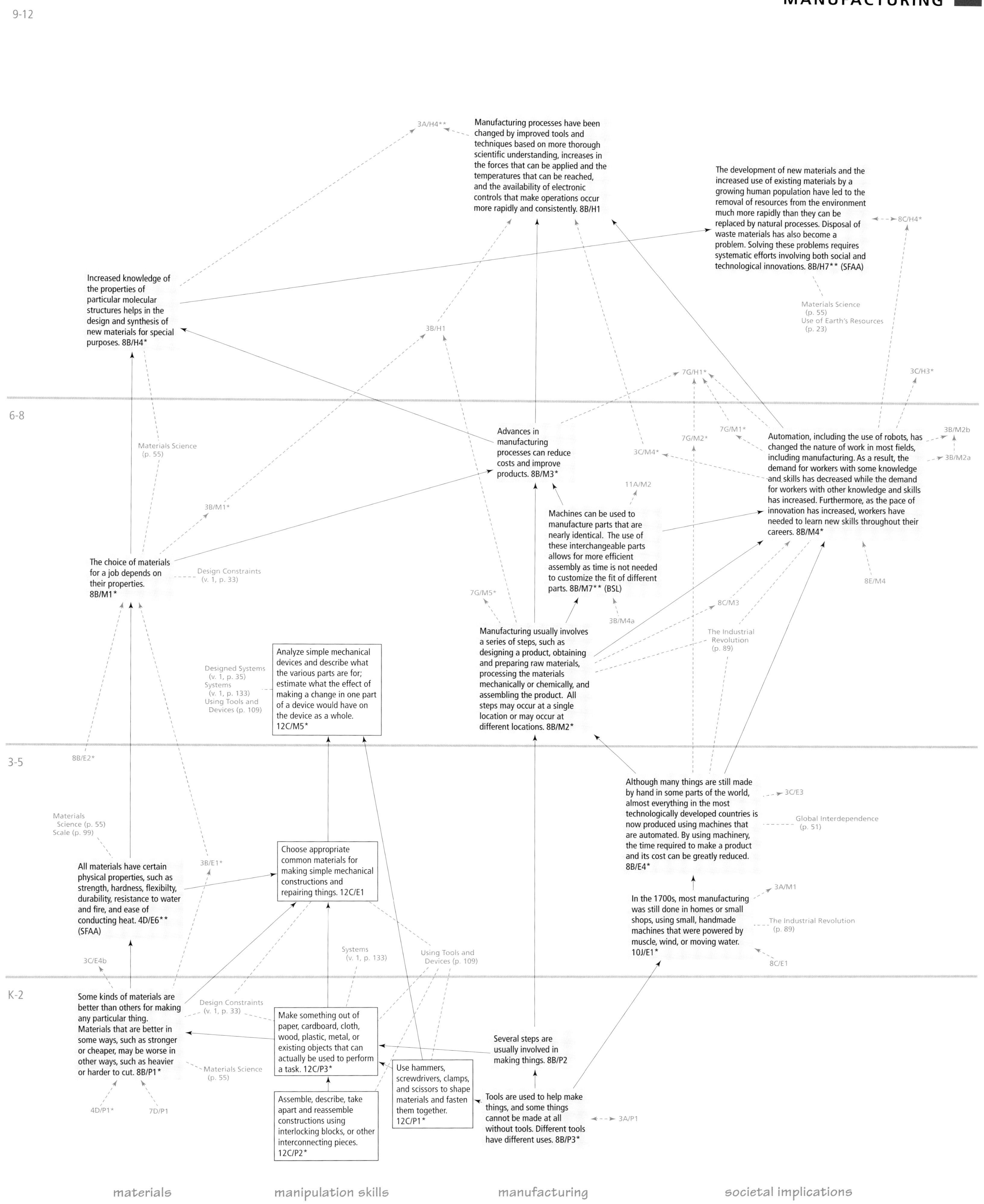
9-12
6-8
3-5
K-2
3A/H4**
Manufacturing processes have been changed by improved tools and techniques based on more thorough scientific understanding, increases in the forces that can be applied and the temperatures that can be reached, and the availability of electronic controls that make operations occur more rapidly and consistently. 8B/H1
The development of new materials and the increased use of existing materials by a growing human population have led to the removal of resources from the environment much more rapidly than they can be replaced by natural processes. Disposal of waste materials has also become a problem. Solving these problems requires systematic efforts involving both social and technological innovations. 8B/H7** (SFAA)
8C/H4*
Increased knowledge of the properties of particular molecular structures helps in the design and synthesis of new materials for special purposes. 8B/H4*
3B/H1
Materials Science (p. 55)
Use of Earth's Resources (p. 23)
7G/H1*
3C/H3*
Materials Science (p. 55)
Advances in manufacturing processes can reduce costs and improve products. 8B/M3*
7G/M2*
7G/M1*
3C/M4*
Automation, including the use of robots, has changed the nature of work in most fields, including manufacturing. As a result, the demand for workers with some knowledge and skills has decreased while the demand for workers with other knowledge and skills has increased. Furthermore, as the pace of innovation has increased, workers have needed to learn new skills throughout their careers. 8B/M4*
3B/M2b
3B/M2a
11A/M2
3B/M1*
Machines can be used to manufacture parts that are nearly identical. The use of these interchangeable parts allows for more efficient assembly as time is not needed to customize the fit of different parts. 8B/M7** (BSL)
The choice of materials for a job depends on their properties. 8B/M1*
Design Constraints (v. 1, p. 33)
7G/M5*
3B/M4a
8E/M4
8C/M3
Analyze simple mechanical devices and describe what the various parts are for; estimate what the effect of making a change in one part of a device would have on the device as a whole. 12C/M5*
Designed Systems (v. 1, p. 35)
Systems (v. 1, p. 133)
Using Tools and Devices (p. 109)
Manufacturing usually involves a series of steps, such as designing a product, obtaining and preparing raw materials, processing the materials mechanically or chemically, and assembling the product. All steps may occur at a single location or may occur at different locations. 8B/M2*
The Industrial Revolution (p. 89)
8B/E2*
Although many things are still made by hand in some parts of the world, almost everything in the most technologically developed countries is now produced using machines that are automated. By using machinery, the time required to make a product and its cost can be greatly reduced. 8B/E4*
3C/E3
Global Interdependence (p. 51)
Materials Science (p. 55)
Scale (p. 99)
3B/E1*
All materials have certain physical properties, such as strength, hardness, flexibilty, durability, resistance to water and fire, and ease of conducting heat. 4D/E6** (SFAA)
Choose appropriate common materials for making simple mechanical constructions and repairing things. 12C/E1
In the 1700s, most manufacturing was still done in homes or small shops, using small, handmade machines that were powered by muscle, wind, or moving water. 10J/E1*
3A/M1
The Industrial Revolution (p. 89)
8C/E1
3C/E4b
Systems (v. 1, p. 133)
Using Tools and Devices (p. 109)
Some kinds of materials are better than others for making any particular thing. Materials that are better in some ways, such as stronger or cheaper, may be worse in other ways, such as heavier or harder to cut. 8B/P1*
Design Constraints (v. 1, p. 33)
Make something out of paper, cardboard, cloth, wood, plastic, metal, or existing objects that can actually be used to perform a task. 12C/P3*
Several steps are usually involved in making things. 8B/P2
Materials Science (p. 55)
Use hammers, screwdrivers, clamps, and scissors to shape materials and fasten them together. 12C/P1*
Tools are used to help make things, and some things cannot be made at all without tools. Different tools have different uses. 8B/P3*
3A/P1
4D/P1*
7D/P1
Assemble, describe, take apart and reassemble constructions using interlocking blocks, or other interconnecting pieces. 12C/P2*
materials
manipulation skills
manufacturing
societal implications

# ENERGY RESOURCES (8C)

Industry, transportation, urban development, agriculture, and most other human activities are closely tied to the amount and kind of energy available. Energy is required for technological processes and needs to be distributed from its source to where it is to be used. Energy sources and ways of using them have different costs, implications and risks. Moreover, the increasing world demand for energy raises national and global issues and the need for an informed public to address them.

The map is organized around three strands—*resources, efficient use,* and *societal and environmental implications*. In the elementary grades the focus is on energy resources and basic ideas about heat transfer. In middle school the emphasis shifts to the transformation and distribution of various energy resources and their environmental consequences. In high school the focus is on trade-offs of various energy resources, the efficiency of energy transformations, and the inevitable dissipation of thermal energy into the environment. At each level, benchmarks about energy are connected to benchmarks about technology and societal implications of energy use.

This map is closely related to the **ENERGY TRANSFORMATIONS** and **THE INDUSTRIAL REVOLUTION** maps. It also illustrates general principles about the nature of technology in **THE INTERACTION OF TECHNOLOGY AND SOCIETY** and **DECISIONS ABOUT USING TECHNOLOGY** maps in *Atlas 1* and the **SCIENCE AND SOCIETY** and **TECHNOLOGY AND SCIENCE** maps in this volume and can serve as a context for learning benchmarks on these maps.

## NOTES

The *resources* strand moves from ideas about the variety of energy resources available to ideas about whether different resources are renewable or not and why. The benchmark "Sunlight is the ultimate source..." is positioned at the 3-5 grade level in *Benchmarks* but has been delayed until the 9-12 level in the map. This change in grade level placement recognizes that understanding how the energy in fossil fuels comes from energy captured by plants long ago requires students to have a more sophisticated knowledge of matter and energy transformations in ecosystems and of processes that shape the earth.

In the *efficient use* strand, the new grades 9-12 benchmark "The useful energy output of a device..." is to be interpreted broadly and intends for students to understand the concept of designing devices to maximize the efficiency of transformation. A prerequisite to understanding the concept of "useful energy" is the idea expressed in benchmark 4E/H3 that less can be done with energy that is spread out.

Two new middle-school benchmarks in the *efficient use* strand come from *Science for All Americans*. The benchmark "Energy is required for technological processes..." describes a variety of processes that use energy and provides the background for appreciating the increased demand and use of energy in industrialized societies. Another new benchmark 8C/M8 expects students to become familiar with a variety of energy-transforming devices and consider what their inputs and outputs are in preparation for discussing the devices' efficiency.

Energy use is always associated with costs and benefits. Hence, the high-school benchmark "Decisions to slow the depletion of energy resources..." is linked to a number of benchmarks related to considering costs and benefits of using technologies, the complexities of decision making, and social trade-offs.

## RESEARCH IN *BENCHMARKS*

Students may identify the transfer of energy to the environment with pollution or waste materials being thrown to the environment (Amettler & Pinto, 2002). Students are not likely to be aware of the heating of the environment which accounts for the energy dissipated (Driver et al., 1994b). Students who have developed an awareness of heat transfer to the environment as a component of most energy transformations may assume that such a transfer results in energy "being lost." They do not think that energy has become spread out and is not as useful as energy which is "concentrated" in one place (Driver et al., 1994b).

Middle-school students do not always explain the process of heating and cooling in terms of heat transfer (Tiberghien, 1983; Tomasini & Balandi, 1987). Some students think that "cold" is being transferred from a colder to a warmer object, others that both "heat" and "cold" are transferred at the same time. Middle- and high-school students often do not explain heat conduction phenomena as interactions (Kesidou, Duit, & Glynn, 1995; Tiberghien, 1985). For example, students sometimes think that, depending on its properties, an object which is at the same temperature with its surroundings can cool down spontaneously. Even after instruction, students don't always give up their naive notion that some substances (for example, flour, sugar, or air) cannot heat up (Tiberghien, 1985) or that metals get hot quickly because "they attract heat," "suck heat in," or "hold heat well" (Erickson, 1985; Lewis & Linn, 1994). Middle-school students believe different materials in the same surroundings have different temperatures because they feel different (for example, metal feels colder than wood). As a result, they may not recognize the universal tendency to temperature equalization (Tomasini & Balandi, 1987; Lewis & Linn, 1994). Helping most students develop scientific ideas in this domain appears to be possible only with long-term interventions that focus on connecting scientific ideas about rate of heat transfer, insulation/conduction, and thermal equilibrium with each other and with ideas students hold based on everyday experiences (Clark & Linn, 2003; Clark & Jorde, 2004; Clark, 2006).

See **ENERGY TRANSFORMATIONS** for additional research.

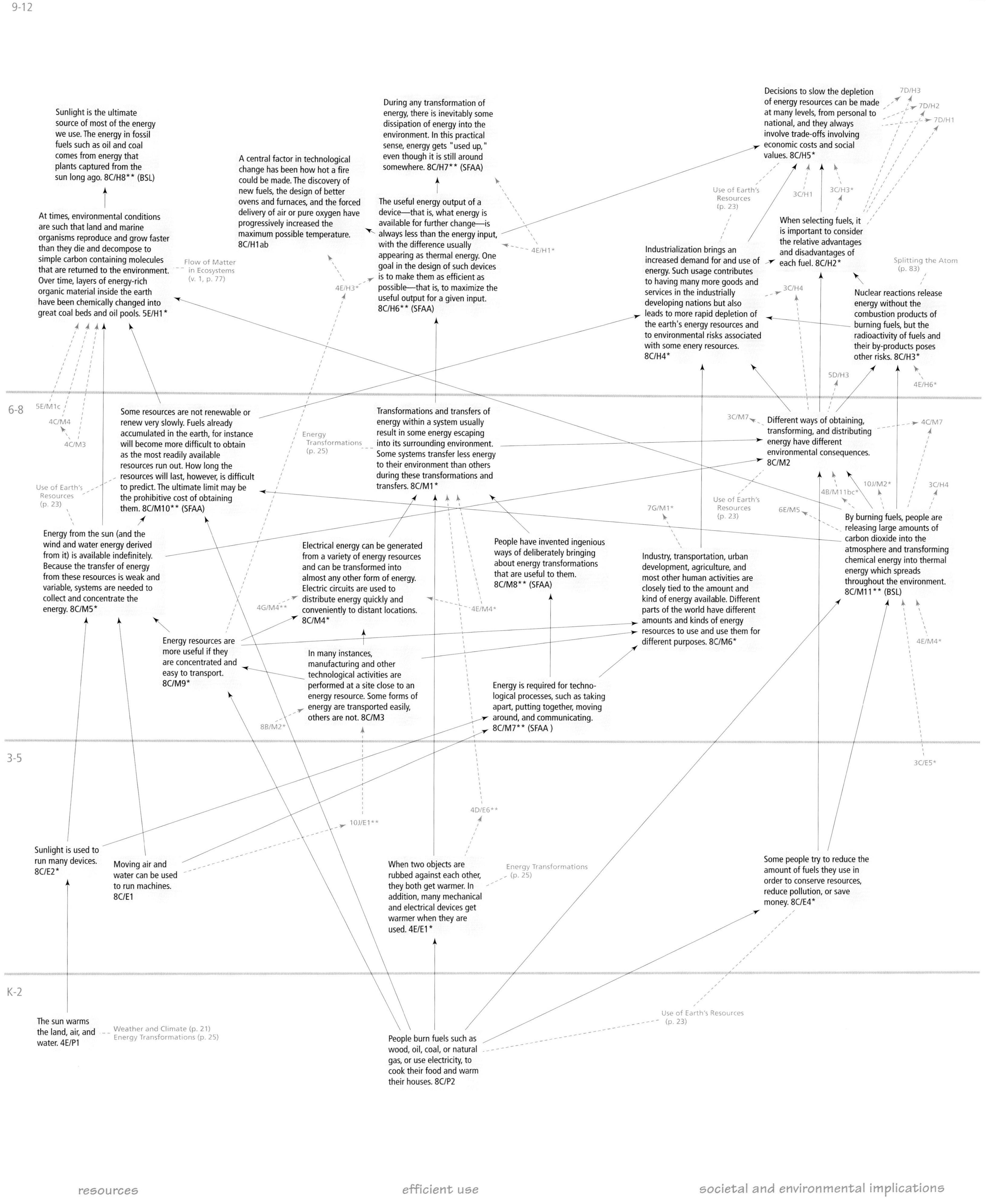
9-12
Sunlight is the ultimate source of most of the energy we use. The energy in fossil fuels such as oil and coal comes from energy that plants captured from the sun long ago. 8C/H8** (BSL)
At times, environmental conditions are such that land and marine organisms reproduce and grow faster than they die and decompose to simple carbon containing molecules that are returned to the environment. Over time, layers of energy-rich organic material inside the earth have been chemically changed into great coal beds and oil pools. 5E/H1*
Flow of Matter in Ecosystems (v. 1, p. 77)
A central factor in technological change has been how hot a fire could be made. The discovery of new fuels, the design of better ovens and furnaces, and the forced delivery of air or pure oxygen have progressively increased the maximum possible temperature. 8C/H1ab
During any transformation of energy, there is inevitably some dissipation of energy into the environment. In this practical sense, energy gets "used up," even though it is still around somewhere. 8C/H7** (SFAA)
The useful energy output of a device—that is, what energy is available for further change—is always less than the energy input, with the difference usually appearing as thermal energy. One goal in the design of such devices is to make them as efficient as possible—that is, to maximize the useful output for a given input. 8C/H6** (SFAA)
4E/H3*
4E/H1*
Decisions to slow the depletion of energy resources can be made at many levels, from personal to national, and they always involve trade-offs involving economic costs and social values. 8C/H5*
7D/H3
7D/H2
7D/H1
Use of Earth's Resources (p. 23)
3C/H1
3C/H3*
Industrialization brings an increased demand for and use of energy. Such usage contributes to having many more goods and services in the industrially developing nations but also leads to more rapid depletion of the earth's energy resources and to environmental risks associated with some enery resources. 8C/H4*
When selecting fuels, it is important to consider the relative advantages and disadvantages of each fuel. 8C/H2*
Splitting the Atom (p. 83)
3C/H4
Nuclear reactions release energy without the combustion products of burning fuels, but the radioactivity of fuels and their by-products poses other risks. 8C/H3*
5D/H3
4E/H6*
6-8
5E/M1c
4C/M4
4C/M3
Some resources are not renewable or renew very slowly. Fuels already accumulated in the earth, for instance will become more difficult to obtain as the most readily available resources run out. How long the resources will last, however, is difficult to predict. The ultimate limit may be the prohibitive cost of obtaining them. 8C/M10** (SFAA)
Use of Earth's Resources (p. 23)
Energy Transformations (p. 25)
Transformations and transfers of energy within a system usually result in some energy escaping into its surrounding environment. Some systems transfer less energy to their environment than others during these transformations and transfers. 8C/M1*
3C/M7
Different ways of obtaining, transforming, and distributing energy have different environmental consequences. 8C/M2
4C/M7
4B/M11bc*
10J/M2*
3C/H4
Energy from the sun (and the wind and water energy derived from it) is available indefinitely. Because the transfer of energy from these resources is weak and variable, systems are needed to collect and concentrate the energy. 8C/M5*
Electrical energy can be generated from a variety of energy resources and can be transformed into almost any other form of energy. Electric circuits are used to distribute energy quickly and conveniently to distant locations. 8C/M4*
4G/M4**
4E/M4*
People have invented ingenious ways of deliberately bringing about energy transformations that are useful to them. 8C/M8** (SFAA)
7G/M1*
Use of Earth's Resources (p. 23)
6E/M5
Industry, transportation, urban development, agriculture, and most other human activities are closely tied to the amount and kind of energy available. Different parts of the world have different amounts and kinds of energy resources to use and use them for different purposes. 8C/M6*
By burning fuels, people are releasing large amounts of carbon dioxide into the atmosphere and transforming chemical energy into thermal energy which spreads throughout the environment. 8C/M11** (BSL)
4E/M4*
Energy resources are more useful if they are concentrated and easy to transport. 8C/M9*
In many instances, manufacturing and other technological activities are performed at a site close to an energy resource. Some forms of energy are transported easily, others are not. 8C/M3
8B/M2*
Energy is required for technological processes, such as taking apart, putting together, moving around, and communicating. 8C/M7** (SFAA )
3-5
3C/E5*
10J/E1**
4D/E6**
Sunlight is used to run many devices. 8C/E2*
Moving air and water can be used to run machines. 8C/E1
When two objects are rubbed against each other, they both get warmer. In addition, many mechanical and electrical devices get warmer when they are used. 4E/E1*
Energy Transformations (p. 25)
Some people try to reduce the amount of fuels they use in order to conserve resources, reduce pollution, or save money. 8C/E4*
K-2
The sun warms the land, air, and water. 4E/P1
Weather and Climate (p. 21)
Energy Transformations (p. 25)
Use of Earth's Resources (p. 23)
People burn fuels such as wood, oil, coal, or natural gas, or use electricity, to cook their food and warm their houses. 8C/P2
resources
efficient use
societal and environmental implications

# HEALTH TECHNOLOGY (8F)

Health technology is concerned with reducing the exposure of humans to conditions that threaten health, increasing the body's resistance to such conditions, and minimizing the deleterious effects that do occur. New medical techniques, efficient health care and delivery systems, improved sanitation, and a fuller understanding of the nature of disease give people a better chance of staying healthy than their forebears had. The options opened up by technology require informed decisions by individuals and society.

The map is organized around three strands that reflect the impact of technology on human health—*diagnosis, treatment,* and *prevention*. Benchmarks in the elementary grades focus on health and ways of protecting it. Benchmarks in middle grades emphasize explanations of how the body works, what causes diseases, how diseases are transmitted, and how the body protects itself from them. Benchmarks in high school build on students' increasing understanding of molecules to explain health and disease at the molecular level. At each grade level, ideas about health are connected to relevant aspects of technology.

Understanding the complex and interrelated factors involved in making decisions about health technology requires that students develop ideas about costs and benefits, trade-offs, side effects, and different interests and value judgments in society. Hence, this map is closely related to the **SOCIAL DECISIONS** and **DECISIONS ABOUT USING TECHNOLOGY** maps in *Atlas 1*.

## NOTES

The *prevention* strand consists of two parallel threads—one presenting ideas about sanitation measures and the other about ways the human body defends itself against infectious disease. Both threads are directly related to the germ theory of disease that is the focus of the **DISEASE** map in *Atlas 1* and the historical account of the discovery of germs in the **DISCOVERING GERMS** map.

Ideas about genetic diseases and treatments are not introduced until grades 9-12; students need a well-developed understanding of the basics of genetics before these benchmarks can be addressed. At the 9-12 level, the map points to the science that underlies health technologies, such as molecular chemistry, which makes possible the synthesis of new drugs, and the behavior of waves in materials, which make possible a variety of imaging techniques.

**RESEARCH IN *BENCHMARKS***

No relevant research available in *Benchmarks*.

# HEALTH TECHNOLOGY

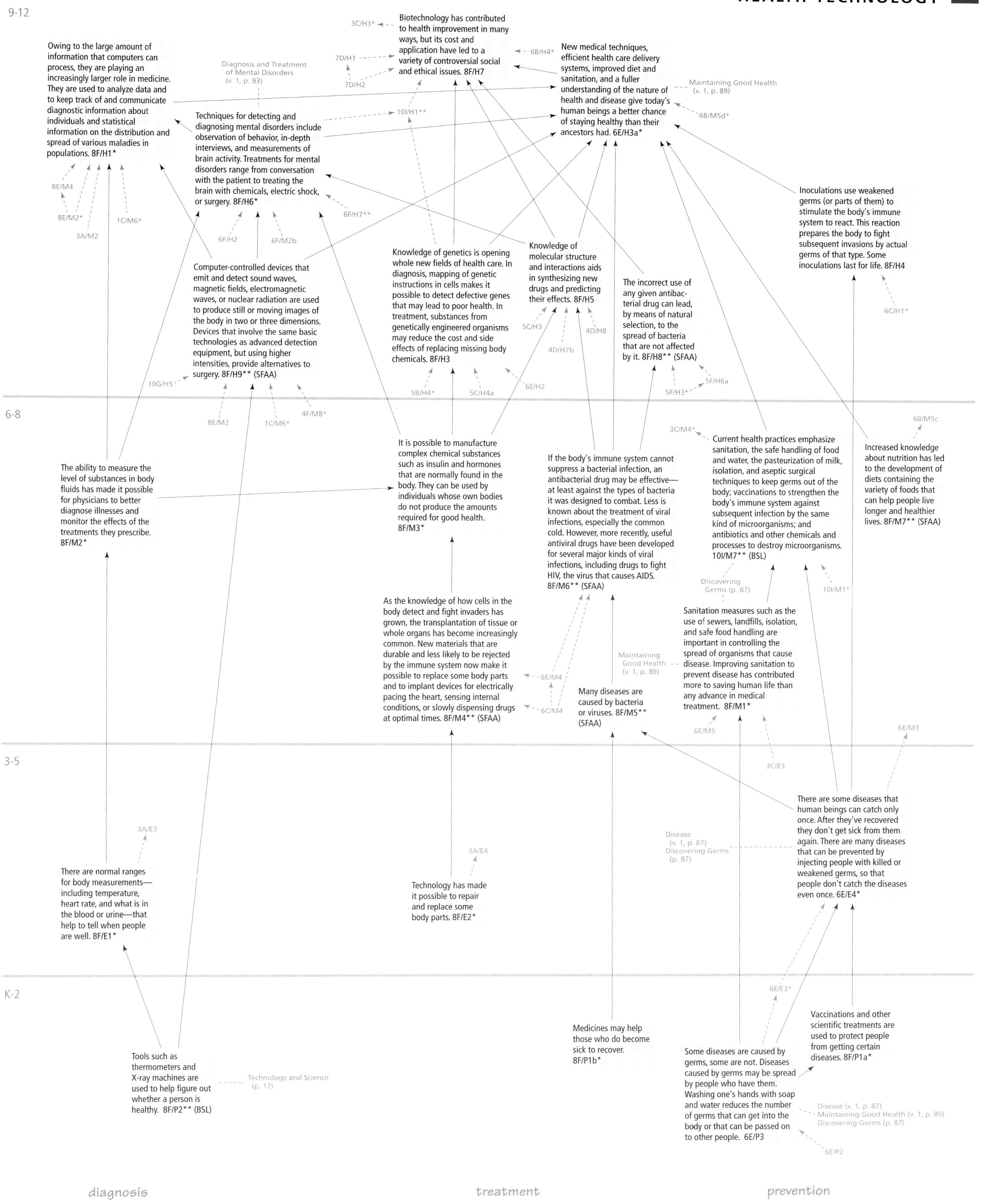

Stuart Davis *Rapt at Rappaport's*, 1952

*Mathematics is essentially a process of thinking that involves building and applying abstract, logically connected networks of ideas. These ideas often arise from the need to solve problems in science, technology, and everyday life—problems ranging from how to model certain aspects of a complex scientific problem to how to balance a checkbook.*

Science for All Americans

# CHAPTER 9



# THE MATHEMATICAL WORLD

The maps in this chapter portray the substance of mathematical literacy and lay out a progression of mathematical ideas students should acquire. In addition, *Atlas of Science Literacy* presents maps dealing with mathematics topics in three other chapters. Chapter 2: THE NATURE OF MATHEMATICS outlines the development of ideas about mathematics as a unique endeavor; Chapter 11: COMMON THEMES lays out a sequence of powerful mathematical concepts, such as scale and models, that are useful as analytical tools across disciplines; and Chapter 12: HABITS OF MIND focuses on the mathematical skills students need, along with scientific and technological skills, to deal effectively with practical affairs in everyday life. Similarly, the maps in this chapter include a great many skill benchmarks as well, emphasizing our contention that skill devoid of understanding, or knowledge that is not grounded in experience and ability, are insufficient for literacy in mathematics.

The list on the left of this page shows the maps for this chapter and indicates the volume of *Atlas of Science Literacy* in which each appears. Those in green typeface are in *Volume 2* and those in gray are in *Volume 1*.

# NUMBERS (9A)

There are several kinds of numbers that, in combination with logic for interrelating them, form interesting abstract systems and can be useful in different ways. The goal is fluency with numbers in describing quantities and the ability to judge whether mathematical thinking makes sense and mathematical results are reasonable—what some refer to as "number sense."

The map is organized around two strands—*number operations* and *describing quantities*. In the elementary grades, the focus is on the meaning of numbers and the ability to compute mentally, on paper, and with a calculator. In middle school, the emphasis is on various ways of representing numbers and on the relationship between various operations on numbers. High school adds fluency with the use of ratio and proportion and powers of ten in solving real-world problems and the recognition that number systems can be constructed with bases other than ten.

This map both draws upon and contributes to ideas on the **COMPUTATION AND ESTIMATION** map.

## NOTES

The equal prominence of knowledge and skills benchmarks on the *number operations* strand illustrates the interdependence of knowing about numbers and operations and having the skills to use them. The grades K-2 benchmark "An important kind of relationship..." clarifies the importance of thinking about parts and wholes and was developed for the **RATIOS AND PROPORTIONS** map in *Atlas 1* as a precursor to ideas about fractions. The new 3-5 benchmark "Symbols are used..." highlights the role of symbols in representing number operations.

Two central benchmarks in the *describing quantities* strand provide the foundation for most of the benchmarks that follow them. The grades K-2 benchmark "Numbers can be used to..." focuses on the role of numbers in describing quantities, and the 6-8 benchmark "The same number can be written in different forms..." ties together students' knowledge about fractions and decimals and paves the way for understanding other number systems. The new K-2 benchmark "The first digit of a two-digit number..." provides an appropriate precursor to a more comprehensive understanding of place value in grades 3-5. Similarly, the new K-2 benchmark "A quantity is stated as a number and a label..." provides support to the 3-5 benchmark on the need to specify any quantity by both a number and a unit. The new 6-8 benchmark "Exponents can be used to represent how many times a number is to be multiplied by itself..." provides support for understanding powers-of-ten notation.

## RESEARCH IN *BENCHMARKS*

During preschool and elementary-school years, children develop meanings for number words in which sequence, count, and cardinal meanings of number words become increasingly integrated (Fuson et al., 1982; Fuson, 1988). Students' own meanings for number words determine to some extent their strategies for adding and subtracting and the complexity of problems they can solve. Elementary- and middle-school students may have limited ability with place value (Sowder, 1992a). Sowder reports that middle-school students are able to identify the place values of the digits that appear in a number, but they cannot use the knowledge confidently in context (for example, students have trouble determining how many boxes of 100 candy bars could be packed from 48,638 candy bars).

Upper elementary- and middle-school students often do not understand that decimal fractions represent concrete objects that can be measured by units, tenths of units, hundredths of units, and so on (Hiebert, 1992). For example, students have trouble writing decimals for shaded parts of rectangular regions divided into 10 or 100 equal parts (Hiebert & Wearne, 1986). Other students have little understanding of the value represented by each of the digits of a decimal number or know the value of the number is the sum of the value of its digits. Students of all ages have problems choosing the largest or the smallest in a set of decimals with different numbers of digits to the right of the decimal points (Carpenter et al., 1981; Hiebert & Wearne, 1986; Resnick et al., 1989). Upper elementary-school students can establish rich meanings for decimal symbols and do a variety of decimal tasks well after specially designed instruction using base-10 blocks (Wearne & Hiebert, 1988, 1989).

Upper elementary- and middle-school students may exhibit limited understanding of the meaning of fractional number (Kieren, 1992). For example, many 7th-graders do not recognize that $5\frac{1}{4}$ is the same as $5 + \frac{1}{4}$ (Kouba et al., 1988). In addition, elementary-school students may have difficulties perceiving a fraction as a single quantity (Sowder, 1988), but rather see it as a pair of whole numbers. An intuitive basis for developing the concept of fractional number is provided by partitioning (Kieren, 1992) and by seeing fractions as multiples of basic units—for example, $\frac{3}{4}$ is $\frac{1}{4}$ and $\frac{1}{4}$ and $\frac{1}{4}$ rather than 3 of 4 parts (Behr et al., 1983).

Middle-school and even high-school students may have limited understanding about the nature and purpose of estimation. They often think it is inferior to exact computation and equate it with guessing (Sowder, 1992b), so that they do not believe estimation is useful (Sowder & Wheeler, 1989). Students who see estimation as a valuable tactic for obtaining information use estimation more frequently and successfully (Threadgill-Sowder, 1984).

There is very little research into student understanding of number symbols as arbitrary conventions. It does indicate that not until 11 years of age do most children consider that correct counting with nonstandard symbols is as adequate as correct counting with standard symbols (Saxe et al., 1989).

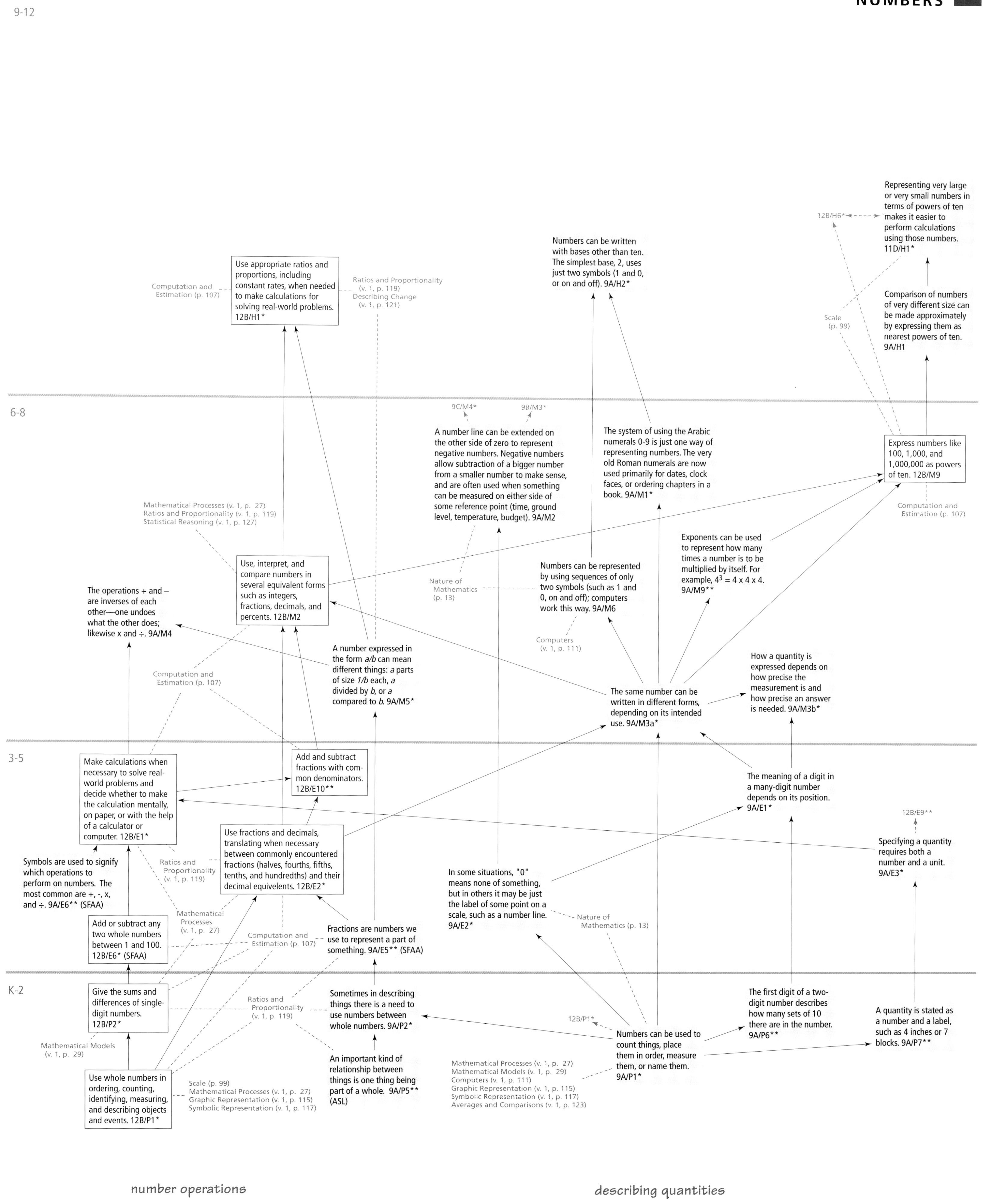
9-12
Use appropriate ratios and proportions, including constant rates, when needed to make calculations for solving real-world problems. 12B/H1*
Computation and Estimation (p. 107)
Ratios and Proportionality (v. 1, p. 119)
Describing Change (v. 1, p. 121)
Numbers can be written with bases other than ten. The simplest base, 2, uses just two symbols (1 and 0, or on and off). 9A/H2*
Representing very large or very small numbers in terms of powers of ten makes it easier to perform calculations using those numbers. 11D/H1*
12B/H6*
Comparison of numbers of very different size can be made approximately by expressing them as nearest powers of ten. 9A/H1
Scale (p. 99)
6-8
9C/M4*
9B/M3*
A number line can be extended on the other side of zero to represent negative numbers. Negative numbers allow subtraction of a bigger number from a smaller number to make sense, and are often used when something can be measured on either side of some reference point (time, ground level, temperature, budget). 9A/M2
The system of using the Arabic numerals 0-9 is just one way of representing numbers. The very old Roman numerals are now used primarily for dates, clock faces, or ordering chapters in a book. 9A/M1*
Express numbers like 100, 1,000, and 1,000,000 as powers of ten. 12B/M9
Computation and Estimation (p. 107)
Mathematical Processes (v. 1, p. 27)
Ratios and Proportionality (v. 1, p. 119)
Statistical Reasoning (v. 1, p. 127)
Use, interpret, and compare numbers in several equivalent forms such as integers, fractions, decimals, and percents. 12B/M2
The operations + and – are inverses of each other—one undoes what the other does; likewise x and ÷. 9A/M4
Nature of Mathematics (p. 13)
Numbers can be represented by using sequences of only two symbols (such as 1 and 0, on and off); computers work this way. 9A/M6
Exponents can be used to represent how many times a number is to be multiplied by itself. For example, 4³ = 4 x 4 x 4. 9A/M9**
Computers (v. 1, p. 111)
Computation and Estimation (p. 107)
A number expressed in the form a/b can mean different things: a parts of size 1/b each, a divided by b, or a compared to b. 9A/M5*
The same number can be written in different forms, depending on its intended use. 9A/M3a*
How a quantity is expressed depends on how precise the measurement is and how precise an answer is needed. 9A/M3b*
3-5
Make calculations when necessary to solve real-world problems and decide whether to make the calculation mentally, on paper, or with the help of a calculator or computer. 12B/E1*
Add and subtract fractions with common denominators. 12B/E10**
The meaning of a digit in a many-digit number depends on its position. 9A/E1*
12B/E9**
Specifying a quantity requires both a number and a unit. 9A/E3*
Symbols are used to signify which operations to perform on numbers. The most common are +, -, x, and ÷. 9A/E6** (SFAA)
Ratios and Proportionality (v. 1, p. 119)
Use fractions and decimals, translating when necessary between commonly encountered fractions (halves, fourths, fifths, tenths, and hundredths) and their decimal equivelents. 12B/E2*
In some situations, "0" means none of something, but in others it may be just the label of some point on a scale, such as a number line. 9A/E2*
Nature of Mathematics (p. 13)
Mathematical Processes (v. 1, p. 27)
Add or subtract any two whole numbers between 1 and 100. 12B/E6* (SFAA)
Computation and Estimation (p. 107)
Fractions are numbers we use to represent a part of something. 9A/E5** (SFAA)
K-2
Give the sums and differences of single-digit numbers. 12B/P2*
Ratios and Proportionality (v. 1, p. 119)
Sometimes in describing things there is a need to use numbers between whole numbers. 9A/P2*
12B/P1*
Numbers can be used to count things, place them in order, measure them, or name them. 9A/P1*
The first digit of a two-digit number describes how many sets of 10 there are in the number. 9A/P6**
A quantity is stated as a number and a label, such as 4 inches or 7 blocks. 9A/P7**
Mathematical Models (v. 1, p. 29)
Use whole numbers in ordering, counting, identifying, measuring, and describing objects and events. 12B/P1*
Scale (p. 99)
Mathematical Processes (v. 1, p. 27)
Graphic Representation (v. 1, p. 115)
Symbolic Representation (v. 1, p. 117)
An important kind of relationship between things is one thing being part of a whole. 9A/P5** (ASL)
Mathematical Processes (v. 1, p. 27)
Mathematical Models (v. 1, p. 29)
Computers (v. 1, p. 111)
Graphic Representation (v. 1, p. 115)
Symbolic Representation (v. 1, p. 117)
Averages and Comparisons (v. 1, p. 123)
number operations
describing quantities

# SHAPES (9C)

To make sense of the world, the human mind relies heavily on its perception of shapes and patterns. The artifacts around us and the familiar forms we see in nature can often be characterized in terms of geometric shapes. Although real-world objects never perfectly match a geometric figure, they more or less approximate them. As a result, what is known about geometric figures and relationships can be applied to objects.

The map is organized around two strands—*shapes in the real world* and *properties of shapes and their relationships*. In the elementary grades, the focus is on relating real-world objects to familiar geometric shapes. In middle school, the focus is on relationships among different aspects of shapes and between shapes and the properties of objects that resemble them. High school adds symmetry to the aspects of shape that affect properties of objects and the recognition that the relationship between real-world objects and geometric shapes is approximate.

This map both draws upon and contributes to ideas on other maps dealing with mathematics, such as **NATURE OF MATHEMATICS, MODELS, PATTERNS OF CHANGE,** and **COMPUTATION AND ESTIMATION**.

## NOTES

The *properties of shapes and their relationships* strand includes three new benchmarks at the 3-5 grade range. These provide examples of simple relationships between shapes to prepare students for the more formal treatment of geometric relationships in the 6-8 benchmark "Shapes can be compared in terms of...." An earlier version of this benchmark was positioned at the 3-5 grade range in *Benchmarks for Science Literacy* but has been moved to the 6-8 level in the map due to the complexity and level of abstraction of the concepts involved.

A new 6-8 benchmark "Geometric relationships..." links geometric relationships to symbolic equations and ties together examples of relationships described in other 6-8 benchmarks. A broader idea—that geometric shapes and relationships can be described in terms of symbols and numbers (and vice versa)—is dealt with in the **GRAPHIC REPRESENTATION** map in *Atlas 1*.

## RESEARCH IN *BENCHMARKS*

Students advance through levels of thought in geometry. Van Hiele has characterized them as visual, descriptive, abstract/relational, and formal deduction (Van Hiele, 1986; Clements & Battista, 1992). At the first level, students identify shapes and figures according to their concrete examples. For example, a student may say that a figure is a rectangle because it looks like a door. At the second level, students identify shapes according to their properties, and here a student might think of a rhombus as a figure with four equal sides. At the third level, students can identify relationships between classes of figures (e.g., a square is a rectangle) and can discover properties of classes of figures by simple logical deduction. At the fourth level, students can produce a short sequence of statements to logically justify a conclusion and can understand that deduction is the method of establishing geometric truth.

Progress from one of Van Hiele's levels to the next is more dependent upon instruction than age. Given traditional instruction, middle-school students perform at levels one or two (Clements & Battista, 1992). Despite that, almost 40% of high-school graduates finish high-school geometry below level two (Burger & Shaughnessy, 1986; Clements & Battista, 1992; Suydam, 1985). Further research will help identify what levels of geometric thinking students can attain at different grades given effective instruction that takes account of their difficulties in learning geometry. Some evidence suggests it is possible for students to understand the abstract properties of geometric figures by 5th grade (Clements & Battista, 1989, 1990, 1992; Wirszup, 1976) and to understand the relations that connect the properties of shapes or make simple deductions by 8th or 9th grade (Clements & Battista, 1992).

Research on students' development of the ability to construct proofs reflects somewhat conflicting views (Clements & Battista, 1992). Piagetian research suggests that students can reason deductively from any assumptions once they reach the formal operational stage (roughly age 12 and beyond). Other research, however, suggests that the ability to construct proofs depends on the amount and organization of particular knowledge they have. For example, this research indicates that students are not likely to understand and construct geometric proofs before they can see the relationships between classes of figures (Senk, 1989). Still other research suggests that students may need to understand the nature of proof and how it differs from everyday argumentation before they are able to construct proofs (Clements & Battista, 1992). Clearly, further research is needed to identify how students can come to understand what it means to prove something in geometry and what such a proof entails.

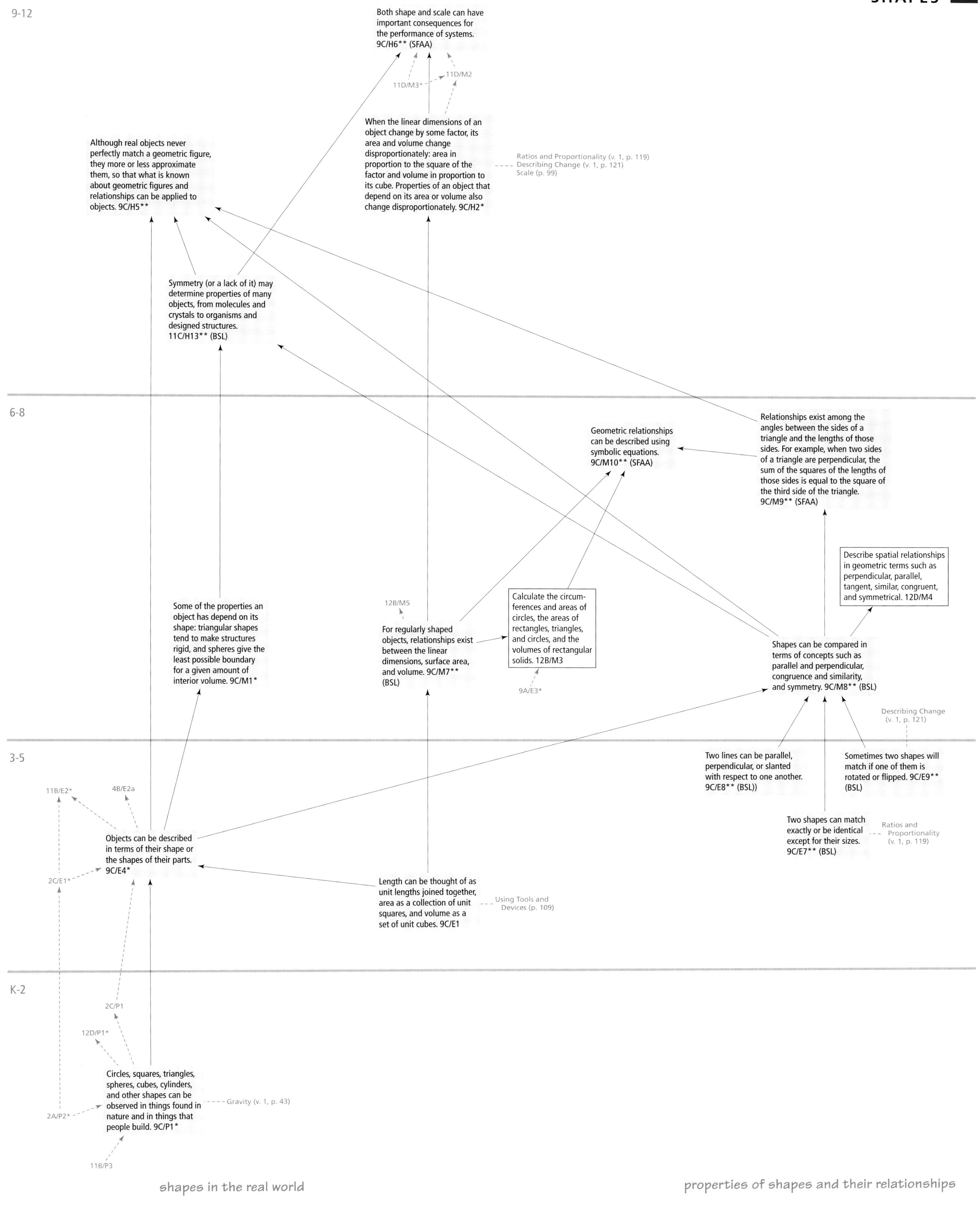
9-12
Both shape and scale can have important consequences for the performance of systems. 9C/H6** (SFAA)
11D/M3*
11D/M2
When the linear dimensions of an object change by some factor, its area and volume change disproportionately: area in proportion to the square of the factor and volume in proportion to its cube. Properties of an object that depend on its area or volume also change disproportionately. 9C/H2*
Ratios and Proportionality (v. 1, p. 119)
Describing Change (v. 1, p. 121)
Scale (p. 99)
Although real objects never perfectly match a geometric figure, they more or less approximate them, so that what is known about geometric figures and relationships can be applied to objects. 9C/H5**
Symmetry (or a lack of it) may determine properties of many objects, from molecules and crystals to organisms and designed structures. 11C/H13** (BSL)
6-8
Geometric relationships can be described using symbolic equations. 9C/M10** (SFAA)
Relationships exist among the angles between the sides of a triangle and the lengths of those sides. For example, when two sides of a triangle are perpendicular, the sum of the squares of the lengths of those sides is equal to the square of the third side of the triangle. 9C/M9** (SFAA)
Describe spatial relationships in geometric terms such as perpendicular, parallel, tangent, similar, congruent, and symmetrical. 12D/M4
Some of the properties an object has depend on its shape: triangular shapes tend to make structures rigid, and spheres give the least possible boundary for a given amount of interior volume. 9C/M1*
12B/M5
For regularly shaped objects, relationships exist between the linear dimensions, surface area, and volume. 9C/M7** (BSL)
Calculate the circumferences and areas of circles, the areas of rectangles, triangles, and circles, and the volumes of rectangular solids. 12B/M3
9A/E3*
Shapes can be compared in terms of concepts such as parallel and perpendicular, congruence and similarity, and symmetry. 9C/M8** (BSL)
Describing Change (v. 1, p. 121)
3-5
Two lines can be parallel, perpendicular, or slanted with respect to one another. 9C/E8** (BSL))
Sometimes two shapes will match if one of them is rotated or flipped. 9C/E9** (BSL)
11B/E2*
4B/E2a
Two shapes can match exactly or be identical except for their sizes. 9C/E7** (BSL)
Ratios and Proportionality (v. 1, p. 119)
Objects can be described in terms of their shape or the shapes of their parts. 9C/E4*
2C/E1*
Length can be thought of as unit lengths joined together, area as a collection of unit squares, and volume as a set of unit cubes. 9C/E1
Using Tools and Devices (p. 109)
K-2
2C/P1
12D/P1*
Circles, squares, triangles, spheres, cubes, cylinders, and other shapes can be observed in things found in nature and in things that people build. 9C/P1*
Gravity (v. 1, p. 43)
2A/P2*
11B/P3
shapes in the real world
properties of shapes and their relationships

# REASONING (9E)

Some aspects of reasoning have clear logical rules; others have only guidelines, and still others have almost unlimited room for creativity. While it is important for students to understand the nature of logic, it is even more important that they learn how to use logic and evidence in making valid, persuasive arguments and in judging the arguments of others.

The map is organized around four strands—*use of deductive logic in reasoning, use of evidence in reasoning, use of inductive logic in reasoning,* and *use of analogies in reasoning*. In the elementary grades, the focus is on backing up claims with reasons. In middle school, the emphasis is on examples of flaws in reasoning. High school culminates with criteria for judging the quality of reasoning.

Ideas on this map draw on and contribute to skills on the **COMMUNICATION SKILLS** and **DETECTING FLAWS IN ARGUMENTS** maps.

## NOTES

Several benchmarks on this map have been revised to either clarify their intent or to improve the coherence of the strands. For example, in the *use of inductive logic in reasoning* strand, the phrase "general rule" was replaced in several benchmarks with the more familiar term "generalization." At the high-school level in this strand, the new benchmark "A failure to find an exception to a generalization…increases the confidence in the accuracy of the generalization" summarizes the use of inductive reasoning.

The new 9-12 benchmark "A sound argument should have both true statements and valid connections among them…" synthesizes ideas about logic and its use and misuse. The term "formal logic" in this benchmark is meant to distinguish between logical arguments that use a strict set of rules in making an argument from those that do not. It is not meant to suggest that an argument is illogical just because the methods of reasoning used in an argument do not adhere to a strict set of rules.

The appearance on this map of many benchmarks dealing with critical response skills highlights the intimate relationship between knowledge about reasoning and the ability to analyze the logic of an argument.

"VERY CREATIVE. VERY IMAGINATIVE. LOGIC... THAT'S WHAT'S MISSING."

## RESEARCH IN *BENCHMARKS*

Students of all ages show a tendency to uncritically infer cause from correlation (Kuhn et al., 1988; Kuhn, 1992). Some students think even a single co-occurrence of antecedent and outcome is always sufficient to infer causality. Rarely do middle-school students realize the indeterminacy of single instances, although high-school students may readily realize it. Despite that, as covariant data accumulate, even high-school students will infer a causal relation based on correlations (Kuhn et al., 1988; Zimmerman, 2000). Further, students of all ages will make a causal inference even when no variation occurs in one of the variables (Kuhn et al., 1988).

Students seem to make valid judgments about situations in which variables have an effect on the outcome earlier than in situations in which variables have no effect on the outcome, or in situations in which, given the current evidence, it is not possible to decide about a variable's role (Schauble, 1996; Kanari & Millar, 2004). Faced with inconclusive data, students may draw conclusions in line with their predictions (Kanari & Millar, 2004). Faced with no correlation of antecedent and outcome, 6th-graders only rarely conclude that the variable has no effect on the outcome (Kuhn et al., 1988). Ninth-graders draw such conclusions more often. A basic problem appears to be understanding the distinction between a variable making no difference and a variable that is correlated with the outcome in the opposite way than the students initially conceived (Kuhn et al., 1988). Another issue is that students are often not aware that all measurements are inevitably subject to uncertainty (or error) and that two measurements of a quantity that has not actually altered are therefore likely to differ (Kanari & Millar, 2004).

A challenge for students of all ages is to generate and interpret evidence that is inconsistent with their prior beliefs (Schauble, 1996). When data are at odds with students' prior beliefs, experiences, or predictions, students may draw conclusions in line with their prior beliefs or predictions (if their prediction is based on some underlying model of the phenomenon involved) (Amsel & Brock, 1996; Millar & Lubben, 1996). Changing beliefs in response to anomalous data may be impeded primarily because students do not make the correct observations (because of their prior belief) rather than because students ignore, distort, discount the observations, or claim that the observations do not hold in other cases (Chinn & Malhotra, 2002).

Students of all ages tend to consider the effect of only a single quantity on another single quantity (Viennot, 2001). Even in multivariable situations, students tend to consider only one factor as possibly influencing the situation, and as a consequence, may overlook other possible influential factors (Driver et al., 1996). Similarly, students have trouble explaining outcomes that are the additive product of two individual variables and may fluctuate from one variable to another trying to explain which single variable produced the outcome (Kuhn, Black, Keselman, & Kaplan, 2000).

Most high-school students will accept arguments based on inadequate sample size, accept causality from contiguous events, and accept conclusions based on statistically insignificant differences (Jungwirth & Dreyfus, 1990, 1992; Jungwirth, 1987). More students can recognize these inadequacies in arguments after prompting (for example, after being told that the conclusions drawn from the data were invalid and asked to state why) (Jungwirth & Dreyfus, 1992; Jungwirth, 1987).

When constructing or evaluating arguments, the following problems appear in student reasoning: problems with validity of arguments, a naïve conception of argument structure, inappropriate effects of core beliefs on argumentation, inadequate sampling of evidence, and altering the representation of argument and evidence (Driver et al., 2000; Zeidler, 1997).

See **DETECTING FLAWS IN ARGUMENTS** for additional research.

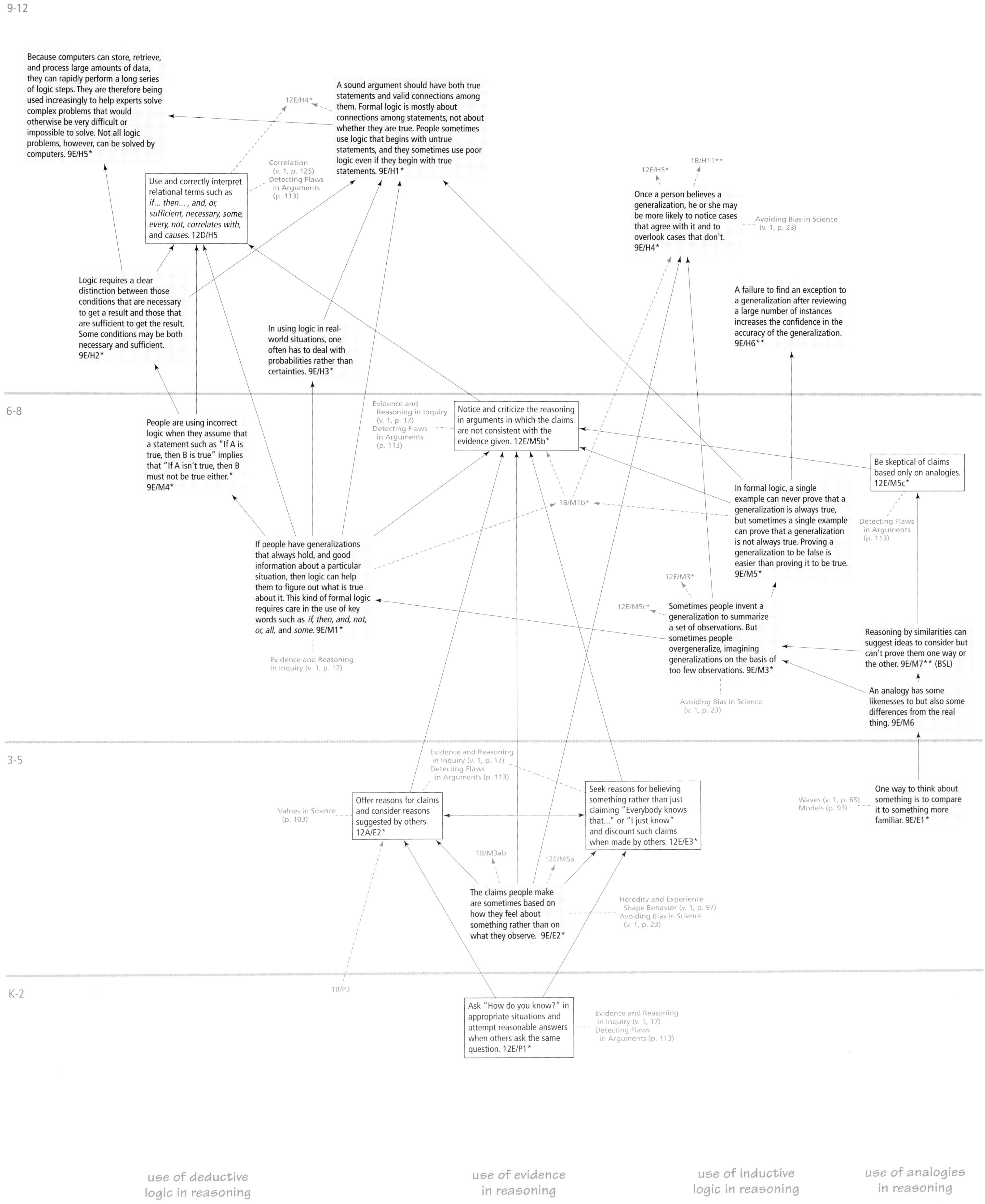
9-12
Because computers can store, retrieve, and process large amounts of data, they can rapidly perform a long series of logic steps. They are therefore being used increasingly to help experts solve complex problems that would otherwise be very difficult or impossible to solve. Not all logic problems, however, can be solved by computers. 9E/H5*
12E/H4*
A sound argument should have both true statements and valid connections among them. Formal logic is mostly about connections among statements, not about whether they are true. People sometimes use logic that begins with untrue statements, and they sometimes use poor logic even if they begin with true statements. 9E/H1*
Correlation (v. 1, p. 125)
Detecting Flaws in Arguments (p. 113)
Use and correctly interpret relational terms such as if... then... , and, or, sufficient, necessary, some, every, not, correlates with, and causes. 12D/H5
12E/H5*
1B/H11**
Once a person believes a generalization, he or she may be more likely to notice cases that agree with it and to overlook cases that don't. 9E/H4*
Avoiding Bias in Science (v. 1, p. 23)
Logic requires a clear distinction between those conditions that are necessary to get a result and those that are sufficient to get the result. Some conditions may be both necessary and sufficient. 9E/H2*
In using logic in real-world situations, one often has to deal with probabilities rather than certainties. 9E/H3*
A failure to find an exception to a generalization after reviewing a large number of instances increases the confidence in the accuracy of the generalization. 9E/H6**
6-8
Evidence and Reasoning in Inquiry (v. 1, p. 17)
Detecting Flaws in Arguments (p. 113)
Notice and criticize the reasoning in arguments in which the claims are not consistent with the evidence given. 12E/M5b*
People are using incorrect logic when they assume that a statement such as "If A is true, then B is true" implies that "If A isn't true, then B must not be true either." 9E/M4*
Be skeptical of claims based only on analogies. 12E/M5c*
1B/M1b*
In formal logic, a single example can never prove that a generalization is always true, but sometimes a single example can prove that a generalization is not always true. Proving a generalization to be false is easier than proving it to be true. 9E/M5*
Detecting Flaws in Arguments (p. 113)
If people have generalizations that always hold, and good information about a particular situation, then logic can help them to figure out what is true about it. This kind of formal logic requires care in the use of key words such as if, then, and, not, or, all, and some. 9E/M1*
12E/M3*
12E/M5c*
Sometimes people invent a generalization to summarize a set of observations. But sometimes people overgeneralize, imagining generalizations on the basis of too few observations. 9E/M3*
Reasoning by similarities can suggest ideas to consider but can't prove them one way or the other. 9E/M7** (BSL)
Evidence and Reasoning in Inquiry (v. 1, p. 17)
Avoiding Bias in Science (v. 1, p. 23)
An analogy has some likenesses to but also some differences from the real thing. 9E/M6
3-5
Evidence and Reasoning in Inquiry (v. 1, p. 17)
Detecting Flaws in Arguments (p. 113)
Values in Science (p. 103)
Offer reasons for claims and consider reasons suggested by others. 12A/E2*
Seek reasons for believing something rather than just claiming "Everybody knows that..." or "I just know" and discount such claims when made by others. 12E/E3*
Waves (v. 1, p. 65)
Models (p. 93)
One way to think about something is to compare it to something more familiar. 9E/E1*
1B/M3ab
12E/M5a
The claims people make are sometimes based on how they feel about something rather than on what they observe. 9E/E2*
Heredity and Experience Shape Behavior (v. 1, p. 97)
Avoiding Bias in Science (v. 1, p. 23)
K-2
1B/P3
Ask "How do you know?" in appropriate situations and attempt reasonable answers when others ask the same question. 12E/P1*
Evidence and Reasoning in Inquiry (v. 1, 17)
Detecting Flaws in Arguments (p. 113)
use of deductive logic in reasoning
use of evidence in reasoning
use of inductive logic in reasoning
use of analogies in reasoning

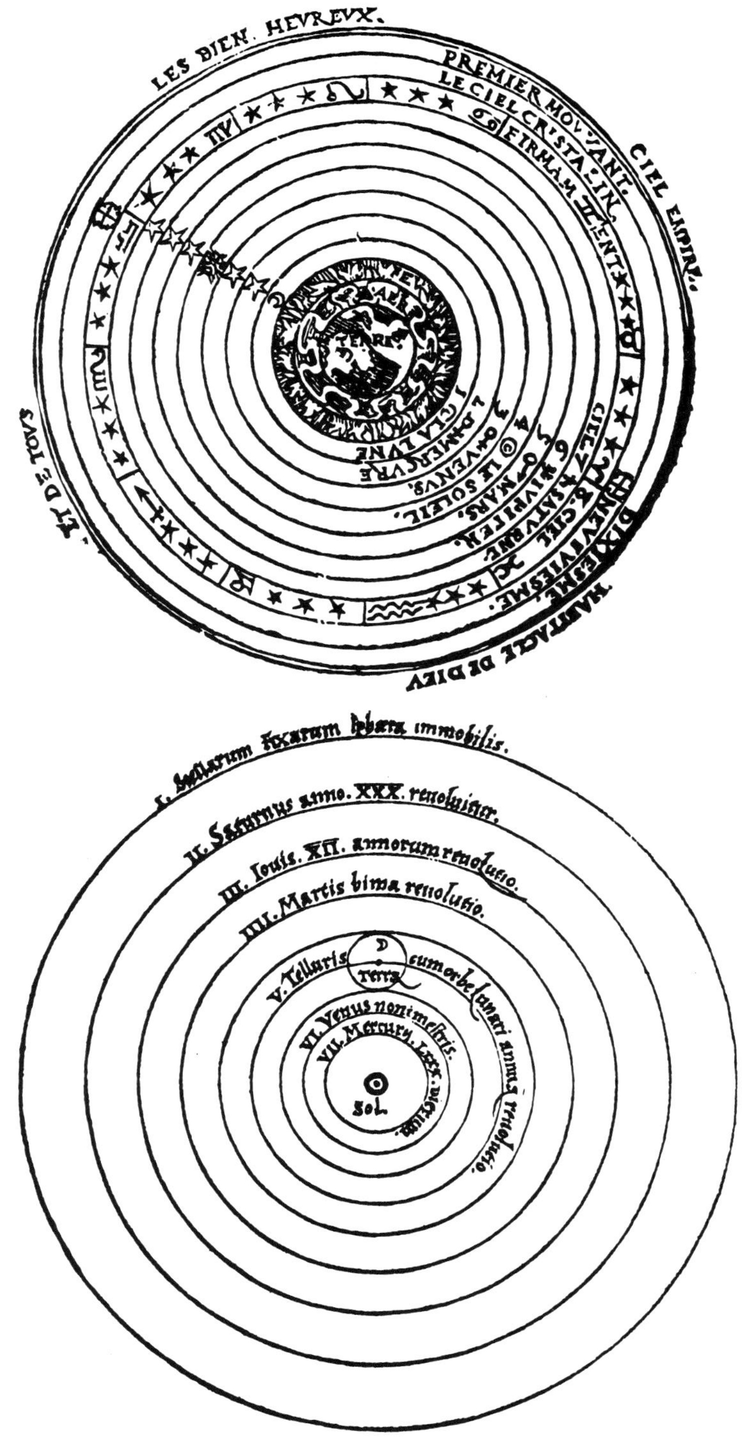


Earth-centered and sun-centered models of the solar system, 16[th] century

*There are two principal reasons for including some knowledge of history among the recommendations. One reason is that generalizations about how the scientific enterprise operates would be empty without concrete examples. Consider, for example, the proposition that new ideas are limited by the context in which they are conceived; are often rejected by the scientific establishment; sometimes spring from unexpected findings; and usually grow slowly, through contributions from many different investigators. Without historical examples, these generalizations would be no more than slogans, however well they might be remembered. For this purpose, any number of episodes might have been selected.*

*A second reason is that some episodes in the history of the scientific endeavor are of surpassing significance to our cultural heritage. Such episodes certainly include Galileo's role in changing our perception of our place in the universe; Newton's demonstration that the same laws apply to motion in the heavens and on earth; Darwin's long observations of the variety and relatedness of life forms that led to his postulating a mechanism for how they came about; Lyell's careful documentation of the unbelievable age of the earth; and Pasteur's identification of infectious disease with tiny organisms that could be seen only with a microscope. These stories stand among the milestones of the development of all thought in Western civilization.*

# CHAPTER 10

# HISTORICAL PERSPECTIVES



The maps in this chapter focus on some important episodes in the history of science, mathematics, and technology. These episodes of scientific discovery and change are important in their own right, but they also contribute to students' understanding of the nature of science and provide a rich context within which students can apply relevant conceptual knowledge. As a result, these maps include benchmarks that present scientific content, historical content, and content related to the nature of science. Each map is accompanied by a timeline that situates the historical episode in the context of other relevant scientific discoveries. Events included in the timelines were selected from *The Timetables of Science* (Hellemans & Bunch, 1991) and from the online resource *Marks in the Evolution of Western Thinking about Nature* (Lee, n.d.) at www.sciencetimeline.net.

There is little research to shed light on teaching and learning these historical topics, so the learning progressions suggested here are more speculative than those for topics in life, earth, and physical science. What research there is tends to be general in nature and does not relate to the specific historical episodes represented on the maps. Therefore, those general findings are summarized below and pertain to all of the maps in this chapter.

The list on the left of this page shows the maps for this chapter and indicates the volume of *Atlas of Science Literacy* in which each appears. All of the maps for Chapter 10 are in Volume 2.

---

## RESEARCH IN *BENCHMARKS*

Research into student understanding of the history of science is limited. Much of the literature consists of descriptions of exemplary practices or untried prescriptions for effective teaching (see, for example, Herget, 1989; Hills, 1992; Matthews, 1991; Shortland & Warwick, 1989). Research in middle-school classrooms has shown that learning some history of science can lead students to a better understanding of the nature of science as well as of the science itself (Solomon et al., 1992).

Research suggests students may have difficulties in understanding the points of view of scientists in the past. Middle-school students show little regard for the thinking of scientists whose theories they know have been superseded (Solomon et al., 1992). Even high-school students have difficulties understanding the points of view of people in the past (Lee, 1984; Shelmit, 1984). In particular, students may think their predecessors were intellectually and morally inferior or may account for their thoughts and behavior with stereotypes before they understand that past values, beliefs, and attitudes were often different from those of today (Shelmit, 1984).

Research into the development of students' broader historical thinking (beyond just the history of science) reflects conflicting views on when history should be taught. On the one hand, some research indicates students are limited in their historical understanding before they reach Piaget's formal-operations stage (Hallam, 1970, 1979; Joyce et al., 1991). Elementary students, for example, have difficulties with time and related aspects like duration and succession (Downey & Levstik, 1988). These results have been used to argue that adolescence is the better time to begin history instruction (Joyce et al., 1991). On the other hand, recent studies indicate that young children know more history facts than has been thought and can think more maturely when they have good background knowledge (Downey & Levstik, 1991). Also, although some children have difficulty with some concepts of time, young children can and do understand historical time in a variety of ways (Egan, 1982; Levstik & Pappas, 1987). They can see patterns and sequences in real events, though some of the patterns may be general and imprecise. These results have been used to argue for an earlier introduction to historical study (Downey & Levstik, 1991). Clearly, more research is needed to assess when and how historical understanding develops in young children and how it can be improved by instruction. Research is also needed to assess whether and how children's conceptions of time are connected to the development of historical understanding.

## HISTORICAL PERSPECTIVES

# THE COPERNICAN REVOLUTION (10A)

The sun-centered model of the universe proposed by Nicolaus Copernicus challenged long-held views of how the world works and of the central importance of the earth and, by extension, of human beings. Indeed, from the perspective of observers standing on the earth, it does appear that the earth is fixed and everything else in the sky moves around it. So along with illustrating important aspects of the way science works, the study of this historical episode highlights an important philosophical question about the relationship between perception and reality.

The map is organized around two strands—*determining the structure of the solar system* and *nature of science*. The progression of understanding begins in elementary and middle school with some fundamental ideas that summarize what is currently known about our solar system. In high school, the focus turns to understanding the historical development of those ideas and their implications for science and for society. The map suggests how this episode can be used to explore the nature of science, the role of technology in scientific discovery, and the ways in which scientific advances can sometimes change people's views of themselves and their world.

Several maps in Chapter 4: THE PHYSICAL SETTING of *Atlas 1*, such as **GALAXIES AND THE UNIVERSE** and **SOLAR SYSTEM**, offer an in-depth treatment of many of the science concepts that are relevant to this map. Other closely linked maps are in Chapter 1: THE NATURE OF SCIENCE in *Atlas 1* and in this volume.

### NOTES

In order to appreciate the historical events surrounding the Copernican Revolution, students need to know about the earth's rotation, its motion around the sun, and how those motions affect what people see when they look up into the sky. It is also important that students have some sense of how large the solar system is to appreciate some of the arguments for and against Copernicus's model of the solar system.

The *determining the structure of the solar system* strand starts with these scientific ideas and then traces their development over time. The ideas are mapped to follow the order in which the relevant scientific discoveries took place. Two new benchmarks appear in this strand. The benchmark "Tycho Brahe, a Danish astronomer..." acknowledges Brahe's contributions, noting that his observations of the planets over many years were more accurate than earlier ones. The other new benchmark "The work of Copernicus, Galileo, Brahe, and Kepler..." points out how the Copernican Revolution changed people's thinking about their place in the universe.

The *nature of science* strand includes benchmarks that describe how new scientific theories are developed and under what conditions they are rejected, accepted, modified, or replaced. Other benchmarks explore how the use of new technologies or the inability to conduct controlled experiments may influence scientific investigations.

### TIMELINE

**300 BCE**

260 BCE Aristarchus of Samos estimates the size of the moon and its distance from Earth.

**100**

100s Ptolemy devises an Earth-centered model of the universe.

**1500**

1543 Nicolaus Copernicus publishes *On the Revolutions of the Heavenly Spheres,* which proposes a sun-centered model of the solar system.

1580 Tycho Brahe begins to collect astronomical data from a specially designed observatory.

1586 Galileo Galilei describes the natural motion of objects as a straight path.

1588 Brahe proposes a system in which the sun orbits the earth, but the other planets revolve around the sun.

1598 Hans Lippershey demonstrates the first telescope.

**1600**

1601 After Brahe's death, Johannes Kepler, who had assisted Brahe, gains access to his astronomical data and begins analyzing it.

1609 Kepler publishes *New Astronomy*, which describes the elliptical orbit of the planets around the sun.

1609 Galileo uses a telescope he built to view Jupiter and discovers four moons orbiting the planet.

1619 Kepler attributes planetary motion to a force exerted by the sun.

1632 Galileo publishes *Dialogue Concerning the Two Chief World Systems,* which favorably compares a sun-centered solar system to an earth-centered one.

1633 Galileo is brought before the Inquisition for publishing *Dialogue Concerning the Two Chief World Systems.*

**1900**

1924 Edwin Hubble concludes that the nearest spiral galaxies are two million light years away, much farther than had been previously thought.

1936 Hubble describes the universe as extending out about 500 million light years (it is now thought to be much larger than that).

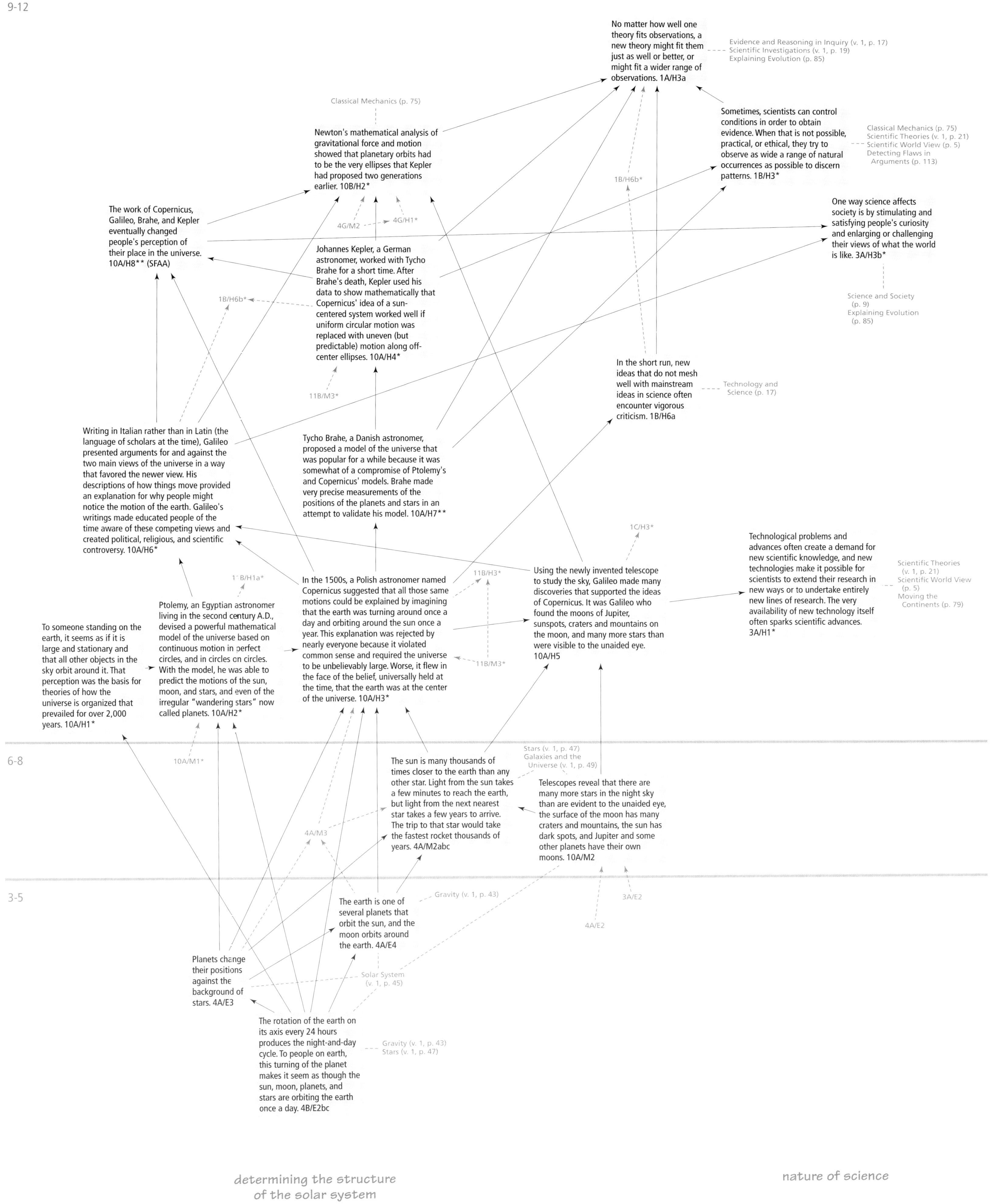
9-12
No matter how well one theory fits observations, a new theory might fit them just as well or better, or might fit a wider range of observations. 1A/H3a
Evidence and Reasoning in Inquiry (v. 1, p. 17)
Scientific Investigations (v. 1, p. 19)
Explaining Evolution (p. 85)
Classical Mechanics (p. 75)
Newton's mathematical analysis of gravitational force and motion showed that planetary orbits had to be the very ellipses that Kepler had proposed two generations earlier. 10B/H2*
Sometimes, scientists can control conditions in order to obtain evidence. When that is not possible, practical, or ethical, they try to observe as wide a range of natural occurrences as possible to discern patterns. 1B/H3*
Classical Mechanics (p. 75)
Scientific Theories (v. 1, p. 21)
Scientific World View (p. 5)
Detecting Flaws in Arguments (p. 113)
1B/H6b*
The work of Copernicus, Galileo, Brahe, and Kepler eventually changed people's perception of their place in the universe. 10A/H8** (SFAA)
4G/M2
4G/H1*
One way science affects society is by stimulating and satisfying people's curiosity and enlarging or challenging their views of what the world is like. 3A/H3b*
Johannes Kepler, a German astronomer, worked with Tycho Brahe for a short time. After Brahe's death, Kepler used his data to show mathematically that Copernicus' idea of a sun-centered system worked well if uniform circular motion was replaced with uneven (but predictable) motion along off-center ellipses. 10A/H4*
1B/H6b*
Science and Society (p. 9)
Explaining Evolution (p. 85)
11B/M3*
In the short run, new ideas that do not mesh well with mainstream ideas in science often encounter vigorous criticism. 1B/H6a
Technology and Science (p. 17)
Writing in Italian rather than in Latin (the language of scholars at the time), Galileo presented arguments for and against the two main views of the universe in a way that favored the newer view. His descriptions of how things move provided an explanation for why people might notice the motion of the earth. Galileo's writings made educated people of the time aware of these competing views and created political, religious, and scientific controversy. 10A/H6*
Tycho Brahe, a Danish astronomer, proposed a model of the universe that was popular for a while because it was somewhat of a compromise of Ptolemy's and Copernicus' models. Brahe made very precise measurements of the positions of the planets and stars in an attempt to validate his model. 10A/H7**
1C/H3*
Technological problems and advances often create a demand for new scientific knowledge, and new technologies make it possible for scientists to extend their research in new ways or to undertake entirely new lines of research. The very availability of new technology itself often sparks scientific advances. 3A/H1*
Scientific Theories (v. 1, p. 21)
Scientific World View (p. 5)
Moving the Continents (p. 79)
1B/H1a*
11B/H3*
Using the newly invented telescope to study the sky, Galileo made many discoveries that supported the ideas of Copernicus. It was Galileo who found the moons of Jupiter, sunspots, craters and mountains on the moon, and many more stars than were visible to the unaided eye. 10A/H5
Ptolemy, an Egyptian astronomer living in the second century A.D., devised a powerful mathematical model of the universe based on continuous motion in perfect circles, and in circles on circles. With the model, he was able to predict the motions of the sun, moon, and stars, and even of the irregular "wandering stars" now called planets. 10A/H2*
In the 1500s, a Polish astronomer named Copernicus suggested that all those same motions could be explained by imagining that the earth was turning around once a day and orbiting around the sun once a year. This explanation was rejected by nearly everyone because it violated common sense and required the universe to be unbelievably large. Worse, it flew in the face of the belief, universally held at the time, that the earth was at the center of the universe. 10A/H3*
To someone standing on the earth, it seems as if it is large and stationary and that all other objects in the sky orbit around it. That perception was the basis for theories of how the universe is organized that prevailed for over 2,000 years. 10A/H1*
11B/M3*
6-8
10A/M1*
Stars (v. 1, p. 47)
Galaxies and the Universe (v. 1, p. 49)
The sun is many thousands of times closer to the earth than any other star. Light from the sun takes a few minutes to reach the earth, but light from the next nearest star takes a few years to arrive. The trip to that star would take the fastest rocket thousands of years. 4A/M2abc
Telescopes reveal that there are many more stars in the night sky than are evident to the unaided eye, the surface of the moon has many craters and mountains, the sun has dark spots, and Jupiter and some other planets have their own moons. 10A/M2
4A/M3
3-5
The earth is one of several planets that orbit the sun, and the moon orbits around the earth. 4A/E4
Gravity (v. 1, p. 43)
3A/E2
4A/E2
Planets change their positions against the background of stars. 4A/E3
Solar System (v. 1, p. 45)
The rotation of the earth on its axis every 24 hours produces the night-and-day cycle. To people on earth, this turning of the planet makes it seem as though the sun, moon, planets, and stars are orbiting the earth once a day. 4B/E2bc
Gravity (v. 1, p. 43)
Stars (v. 1, p. 47)
determining the structure of the solar system
nature of science

## HISTORICAL PERSPECTIVES

# CLASSICAL MECHANICS (10B)

Near the end of the 17th century, in his *Mathematical Principles of Natural Philosophy*, Isaac Newton used just a few key concepts and mathematical relationships to explain the motion of objects on Earth and the motion of distant heavenly bodies. With these few ideas, he could account for the observed orbits of planets and moons, the motion of comets, the irregular motion of the moon, the motion of falling objects at the earth's surface, weight, ocean tides, and the earth's equatorial bulge. But the fact that a simple set of precise mechanical laws could explain so much also raised concerns among philosophers—both within and outside of the scientific community—that perhaps all human activity and thought could be explained by mechanical laws and that free will might be only an illusion.

Classical mechanics continues to be useful for describing the motion of most objects under the conditions that we confront in our everyday lives. The story of the development of classical mechanics illustrates well how mathematics provides a precise language to describe objects and events and the relationships between them. It also illustrates what is now a fundamental assumption in science: that the universe is a vast single system in which the basic rules are the same and that the events in the universe occur in consistent patterns that are comprehensible through careful, systematic study.

The map is organized around two strands—*historical development and influence* and *nature of science*. The learning progression begins with a single middle-school benchmark on gravity; all of the other ideas are introduced in high school.

This map is closely related to other Chapter 10 maps, such as **THE COPERNICAN REVOLUTION** and **RELATIVITY**, and also to maps in Chapter 1: THE NATURE OF SCIENCE and Chapter 4: THE PHYSICAL SETTING.

### NOTES

In order to appreciate the historical significance of classical mechanics, students need to know about gravity and the effects of force on the motion of objects. The *historical development and influence* strand starts with prerequisite benchmarks that describe Isaac Newton's three laws of motion and his law of gravitation. Benchmark 10B/H1 has been revised to better reflect its centrality to the scientific discoveries described on this map, and benchmark 10B/H2 has been modified to focus on the link between Newton's ideas and the motions of the planets that Johannes Kepler had proposed.

In the *nature of science* strand, the focus is on understanding how scientific theories are developed and evaluated; how mathematics provides a way to describe physical phenomena; and how Newton's ideas, and the accurate predictions they made possible, eventually convinced people that the universe could be comprehensible.

### TIMELINE

**1500**

1583 Galileo Galilei discovers that the period of a swinging pendulum is the same regardless of its amplitude.

1586 Galileo describes the natural motion of objects as a straight path.

**1600**

1609 Johannes Kepler publishes *New Astronomy*, which describes the elliptical orbit of the planets around the sun.

1632 Galileo publishes *Dialogue Concerning the Two Chief World Systems*, which favorably compares a sun-centered solar system to that of an earth-centered one.

1638 Galileo publishes *Two New Sciences*, which describes the law of falling bodies.

1666 Isaac Newton begins developing calculus.

1674 Gottfried Leibniz begins developing his own system of calculus, independent of Newton.

1684 Leibniz publishes his system of calculus.

1687 Newton publishes *Principia*, in which he uses his calculus to describe his three laws of motion and the law of gravitation.

**1700**

1704 Newton publishes *Opticks*, describing his work on light.

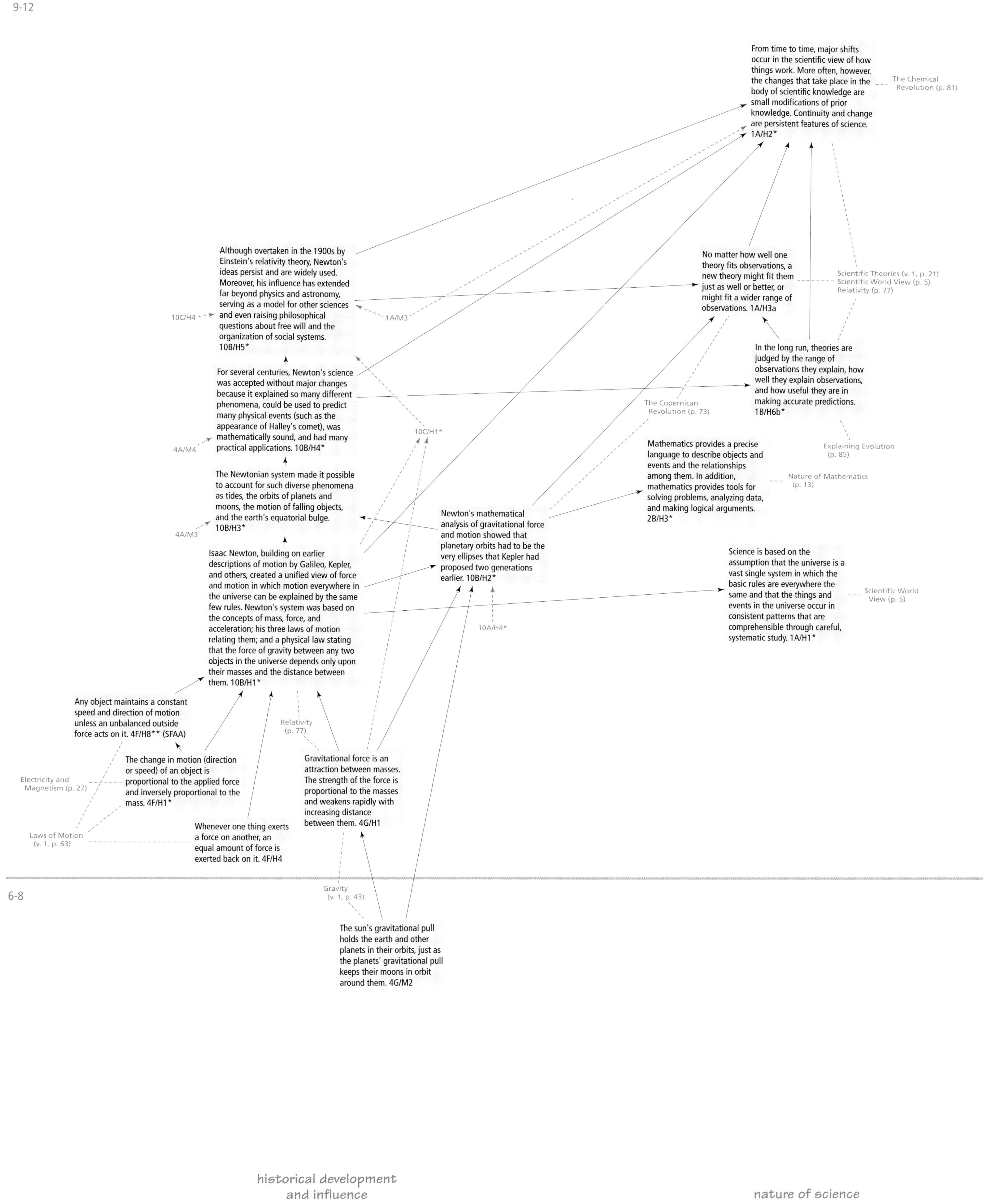
9-12
From time to time, major shifts occur in the scientific view of how things work. More often, however, the changes that take place in the body of scientific knowledge are small modifications of prior knowledge. Continuity and change are persistent features of science. 1A/H2*
The Chemical Revolution (p. 81)
Although overtaken in the 1900s by Einstein's relativity theory, Newton's ideas persist and are widely used. Moreover, his influence has extended far beyond physics and astronomy, serving as a model for other sciences and even raising philosophical questions about free will and the organization of social systems. 10B/H5*
10C/H4
1A/M3
No matter how well one theory fits observations, a new theory might fit them just as well or better, or might fit a wider range of observations. 1A/H3a
Scientific Theories (v. 1, p. 21)
Scientific World View (p. 5)
Relativity (p. 77)
For several centuries, Newton's science was accepted without major changes because it explained so many different phenomena, could be used to predict many physical events (such as the appearance of Halley's comet), was mathematically sound, and had many practical applications. 10B/H4*
4A/M4
In the long run, theories are judged by the range of observations they explain, how well they explain observations, and how useful they are in making accurate predictions. 1B/H6b*
The Copernican Revolution (p. 73)
Explaining Evolution (p. 85)
10C/H1*
Mathematics provides a precise language to describe objects and events and the relationships among them. In addition, mathematics provides tools for solving problems, analyzing data, and making logical arguments. 2B/H3*
Nature of Mathematics (p. 13)
The Newtonian system made it possible to account for such diverse phenomena as tides, the orbits of planets and moons, the motion of falling objects, and the earth's equatorial bulge. 10B/H3*
4A/M3
Newton's mathematical analysis of gravitational force and motion showed that planetary orbits had to be the very ellipses that Kepler had proposed two generations earlier. 10B/H2*
Isaac Newton, building on earlier descriptions of motion by Galileo, Kepler, and others, created a unified view of force and motion in which motion everywhere in the universe can be explained by the same few rules. Newton's system was based on the concepts of mass, force, and acceleration; his three laws of motion relating them; and a physical law stating that the force of gravity between any two objects in the universe depends only upon their masses and the distance between them. 10B/H1*
Science is based on the assumption that the universe is a vast single system in which the basic rules are everywhere the same and that the things and events in the universe occur in consistent patterns that are comprehensible through careful, systematic study. 1A/H1*
Scientific World View (p. 5)
10A/H4*
Any object maintains a constant speed and direction of motion unless an unbalanced outside force acts on it. 4F/H8** (SFAA)
Relativity (p. 77)
The change in motion (direction or speed) of an object is proportional to the applied force and inversely proportional to the mass. 4F/H1*
Electricity and Magnetism (p. 27)
Gravitational force is an attraction between masses. The strength of the force is proportional to the masses and weakens rapidly with increasing distance between them. 4G/H1
Laws of Motion (v. 1, p. 63)
Whenever one thing exerts a force on another, an equal amount of force is exerted back on it. 4F/H4
6-8
Gravity (v. 1, p. 43)
The sun's gravitational pull holds the earth and other planets in their orbits, just as the planets' gravitational pull keeps their moons in orbit around them. 4G/M2
historical development and influence
nature of science

# RELATIVITY (10C)

Around the beginning of the 20th century, Albert Einstein argued that space and time are very closely linked, rather than the completely different dimensions that Isaac Newton had assumed. One of the surprising implications of Einstein's ideas is that the speed of light is the same no matter how fast the person measuring the light or the source of the light is moving; another is that light in empty space moves at the greatest speed possible. His theory of special relativity also asserts the equivalence of mass and energy, expressed as the famous $E=mc^2$. Eventually, Einstein's publication of his theory of general relativity explained Newton's gravitational force as not a force at all but rather as a manifestation of curved space and time—a distortion in the geometry of space and time.

Einstein's ideas have proved to be monumental in the history of science. Because predictions based on his theories of special and general relativity have always been borne out, no matter how counterintuitive they may seem, this episode illustrates well that, in the long run, theories are judged by the range of observations they explain and how effective they are in predicting new findings.

The map is organized around three strands—*mass and energy equivalence, theory development by Einstein*, and *nature of science*. All but one benchmark have been placed at the high-school level.

Closely related maps include other Chapter 10 maps, such as **THE COPERNICAN REVOLUTION** and **CLASSICAL MECHANICS,** as well as maps in Chapter 1: THE NATURE OF SCIENCE in both volumes of *Atlas*.

## NOTES

As a part of basic science literacy, students are asked to know some of the broad conclusions of special and general relativity, such as the equivalence of mass and energy, but are not expected to understand the theory in detail or how the various postulates of the theory lead to the various conclusions. Benchmark 10G/H3b "In proposing that a uranium nucleus could split...," at the top of the *mass and energy equivalence* strand, also appears on the **SPLITTING THE ATOM** map, and several precursor ideas can be found there as well.

Benchmarks 10C/H1 and 10C/H4 in the *theory development by Einstein* strand have been revised to provide clearer explanations of special and general relativity.

The *nature of science* strand uses the development of the theories of special and general relativity to show how one theory often supplants another.

## TIMELINE

**1800**

1887 Albert Michelson and Edward Morley conduct their first interferometer experiments to detect the luminiferous aether (thought to be the medium for light) but find no evidence of it.

**1900**

1905 Albert Einstein writes papers on the photoelectric effect, Brownian motion, special relativity, and matter and energy equivalence.

1907 Einstein first uses the equation $E=mc^2$ to describe the equivalence of mass and energy.

1915 Einstein introduces a theory of general relativity.

1919 British astronomers observe a solar eclipse and note that the bending of starlight is consistent with the predictions of Einstein's theory of general relativity.

1939 Einstein writes President Franklin D. Roosevelt to encourage U.S. research on the possible use of nuclear fission in a weapon.

9-12

From time to time, major shifts occur in the scientific view of how things work. More often, however, the changes that take place in the body of scientific knowledge are small modifications of prior knowledge. Continuity and change are persistent features of science. 1A/H2*

The Chemical Revolution (p. 81)

No matter how well one theory fits observations, a new theory might fit them just as well or better, or might fit a wider range of observations. 1A/H3a

The Copernican Revolution (p. 73)

Scientific World View (p. 5)
Classical Mechanics (p. 75)
Scientific Theories (v. 1, p. 21)

In the long run, theories are judged by the range of observations they explain, how well they explain observations, and how useful they are in making accurate predictions. 1B/H6b*

Explaining Evolution (p. 85)
Moving the Continents (p. 79)

Einstein's development of the theories of special and general relativity ranks as one of the greatest human accomplishments in all of history. Many predictions from the theories have been confirmed on both atomic and astronomical scales. Still, the search continues for an even more powerful theory of the architecture of the universe. 10C/H5*

10G/H3c*

In proposing that a uranium nucleus could split, Meitner had to account for the energy that would be released when the newly formed nuclei accelerated away from each other. She suggested it could come from the conversion of mass to energy predicted by Einstein's theory of special relativity. 10G/H3b*

Splitting the Atom (p. 83)

10G/H3a*

4E/H6*

Under everyday situations, most of the predictions of special relativity are nearly identical to those of classical mechanics. The more counterintuitive predictions of special relativity occur in situations that humans do not typically experience. 10C/H6**

A decade after Einstein developed the special theory of relativity, he proposed the general theory of relativity, which pictures Newton's gravitational force as a distortion of space and time. 10C/H4*

2B/H3*

The special theory of relativity is best known for stating that any form of energy has mass, and that matter itself is a form of energy. Even a tiny amount of matter holds an enormous amount of energy. This relationship is described in the famous relativity equation $E = mc^2$, in which the c in the equation stands for the immense speed of light. 10C/H3*

As a young man, Albert Einstein, a German scientist, formulated the special theory of relativity, which brought about revolutionary changes in human understanding of nature. Among the counterintuitive ideas of special relativity is that the speed of light is the same for all observers no matter how they or the light source happen to be moving. In addition, nothing can travel faster than the speed of light. 10C/H1*

Isaac Newton, building on earlier descriptions of motion by Galileo, Kepler, and others, created a unified view of force and motion in which motion everywhere in the universe can be explained by the same few rules. Newton's system was based on the concepts of mass, force, and acceleration; his three laws of motion relating them; and a physical law stating that the force of gravity between any two objects in the universe depends only upon their masses and the distance between them. 10B/H1*

Waves (v. 1, p. 65)
Galaxies and the Universe (v. 1, p. 49)

In empty space, all electromagnetic waves move at the same speed—the "speed of light." 4F/H3c

Gravitational force is an attraction between masses. The strength of the force is proportional to the masses and weakens rapidly with increasing distance between them. 4G/H1

Gravity (v. 1, p. 43)
Classical Mechanics (p. 75)

6-8

Because every object is moving relative to some other object, no object has a unique claim to be at rest. Therefore, the idea of absolute motion or rest is misleading. 10A/M1*

Gravity (v. 1, p. 43)
Solar System (v. 1, p. 75)
Laws of Motion (v. 1, p. 63)

mass and energy equivalence

theory development by Einstein

nature of science

## HISTORICAL PERSPECTIVES

# MOVING THE CONTINENTS (10DE)

When people began to notice that the continents of Africa and South America have nearly matching coastlines, some speculated that these two continents may once have been attached and then drifted apart. However, the idea that continents could "drift" was difficult to accept because of their enormous size and their apparent immobility. Even as additional evidence supporting the possibility that those continents had once been part of the same landmass began to be presented in the 1960s, scientists were still hesitant to accept the idea of moving continents because they lacked a plausible explanation for that movement. Eventually, the study of the motion of Earth's plates transformed a theory of continental drift into a theory of plate tectonics.

This episode in the history of science illustrates how new ideas that do not mesh well with mainstream ideas encounter vigorous criticism from scientists until a plausible mechanism is offered to account for the phenomena involved.

The map is organized around four strands—*scientific ideas about plate tectonics, development of scientific understanding of the motion of continents, determining the age of the earth,* and *nature of science*. There are no benchmarks prior to middle school, where the focus is on ideas about the earth's plates. High-school benchmarks describe earlier theories and the development of current scientific explanations.

In *Atlas 1*, the **PLATE TECTONICS** map offers more details on the science concepts discussed on this map. In addition, the maps in Chapter 1: THE NATURE OF SCIENCE in both volumes of *Atlas* are relevant.

### NOTES

Several of the benchmarks in the *scientific ideas about plate tectonics* strand are revised versions of the same benchmarks that appear on the **PLATE TECTONICS** map in *Atlas 1*. Revisions have been made as a result of Project 2061's work to clarify ideas on this topic for textbook analysis and assessment design since the publication of *Atlas 1*.

The map presents two related historical episodes in which scientists attempted to answer important questions about the age of the earth and the motion of the continents. They are mapped together to show that knowing that the earth is many millions of years old is essential to knowing that would be sufficient time for some continents to have moved thousands of miles even though their movement is so slow that it requires sensitive equipment to detect.

The 9-12 benchmark "In the 1960s, scientists noted that earthquakes..." has been revised to be more explicit about some of the reasons that the scientific community had become more open to the idea of moving continents than they were when Alfred Wegener first introduced the idea earlier in the century. Benchmark 10E/H4 "Scientists continue to study..." has been added to the map as a reminder that there are many things about plate tectonics that are still not understood.

The *nature of science* strand describes how new ideas in science are evaluated by the scientific community. Like many of the historical episodes mapped in Chapter 10, the events described on this map can be particularly helpful in understanding how scientific theories are developed and tested.

" I WOULDN'T WORRY. WITH CONTINENTAL DRIFT, AFRICA OR SOUTH AMERICA SHOULD COME BY EVENTUALLY. "

### TIMELINE

**1700**

1759 Mikhail Vasilevich Lomonosov suggests that very slow natural activities, such as uplift and erosion, are the cause of Earth's topography.

1795 James Hutton writes about the vast antiquity of geological formations; James Hall reproduces in miniature the processes Hutton claims are responsible for the formation of rock strata under the conditions prevailing in the earth's crust.

**1800**

1830 Charles Lyell publishes the first volume of *Principles of Geology*, presenting evidence that the earth might be much older than the prevailing estimates.

1833 Lyell publishes the third, and final, volume of *Principles of Geology*.

**1900**

1915 Alfred Wegener publishes *The Origin of Continents and Oceans*, proposing a theory of continental drift and the idea that the continents had once been connected as a single landmass, which he later called Pangaea.

1947 Maurice Ewing and other scientists from the Woods Hole Oceanographic Institution find a rise in the seafloor of the central Atlantic Ocean that is eventually known as the Mid-Atlantic Ridge.

1950s Patterns of magnetic variations are discovered in the seafloor.

1960 Harry Hess distributes a paper suggesting that an ocean basin and the adjoining continent move together as a plate.

1967 W. J. Morgan proposes that the surface of the earth consists of 12 rigid plates that move relative to each other.

9-12

Public Perception of Science (p. 105)

In science, the testing, revising, and occasional discarding of theories, new and old, never ends. This ongoing process leads to a better understanding of how things work in the world but not to absolute truth. 1A/H3bc*

The Copernican Revolution (p. 73)
Relativity (p. 77)
Explaining Evolution (p. 85)

In the long run, theories are judged by the range of observations they explain, how well they explain observations, and how useful they are in making accurate predictions. 1B/H6b*

Scientists continue to study the motions of the earth's plates and the phenomena those motions cause in an attempt to better understand the internal composition of the earth and the processes taking place within it. 10E/H4**

In the 1960s, scientists noted that earthquakes occur much more frequently in certain areas, that the rock around mid-ocean ridges is progressively older the farther it is from the ridge, and that this gradient is symmetrical on either side of the ridge. This evidence, coupled with a scientifically sound physical explanation for how continents could move, transformed the idea of moving continents into the theory of plate tectonics. 10E/H3*

New ideas in science are limited by the context in which they are conceived; are often rejected by the scientific establishment; sometimes spring from unexpected findings; and usually grow slowly, through contributions from many investigators. 1B/H7

Scientific World View (p. 5)
Scientific Theories (v. 1, p. 21)

Discovering Germs (p. 87)

In the short run, new ideas that do not mesh well with mainstream ideas in science often encounter vigorous criticism. 1B/H6a

The Copernican Revolution (p. 73)

Early in the 1900s, Alfred Wegener, a German scientist, reintroduced the idea of moving continents, adding such evidence as the underwater shapes of the continents, the similarity of life forms and land forms in corresponding parts of Africa and South America, and the increasing separation of Greenland and Europe. Even with this evidence and the realization that the earth was old enough for this to have occurred, very few contemporary scientists adopted Wegener's theory because he lacked a plausible mechanism for the movement of continents. 10E/H2*

In the early 1800s, Charles Lyell argued in *Principles of Geology* that the earth was vastly older than most people believed. He supported his claim with a wealth of observations of the patterns of rock layers in mountains and the locations of various kinds of fossils. 10D/H2*

Plate Tectonics (v. 1, p. 53)

The outward transfer of the earth's internal heat causes regions of different temperatures and densities. The action of a gravitational force on regions of different densities causes the rise and fall of material between the earth's surface and interior, which is responsible for the movement of plates. 4C/H3*

As soon as fairly accurate world maps began to appear, some people noticed that the continents of Africa and South America looked as though they might fit together, like a giant jigsaw puzzle. This led some to speculate that they might have once been part of a single giant land mass that broke into pieces and then drifted apart. This idea was repeatedly suggested and rejected because it was hard to imagine that anything that large and apparently immobile could move. 10E/H1*

Prior to the 1700s, many considered the earth to be just a few thousand years old. By the 1800s, scientists were starting to realize that the earth was much older even though they could not determine its exact age. 10D/H1*

Explaining Evolution (p. 85)

Scientific evidence indicates that some rock layers are several billion years old. 4C/H6** (BSL)

Changes in the Earth's Surface (v. 1, p. 51)

6-8

There are worldwide patterns to major geological events (such as earthquakes, volcanic eruptions, and mountain building) that coincide with plate boundaries. 4C/M13** (BSL)

4C/M9**

The earth's plates sit on a denser, hot, somewhat melted layer of the earth. The plates move very slowly, pressing against one another in some places and pulling apart in other places—sometimes scraping alongside each other as they do. Mountains form as two continental plates, or an ocean plate and a continental plate, press together. 4C/M12** (BSL)

The outer layer of the earth—including both the continents and the ocean basins—consists of separate plates. 4C/M11** (BSL)

scientific ideas about plate tectonics

development of scientific understanding of the motion of continents

determining the age of the earth

nature of science

# THE CHEMICAL REVOLUTION (10F)

Antoine Lavoisier's idea of conservation of matter became the centerpiece of the modern science of chemistry. By carefully measuring all of the substances involved in burning, including the gases used and given off, Lavoisier was able to show that there is no net gain or loss of weight during burning. Lavoisier also developed a system for naming substances in terms of the elements that make up those substances. This, along with the ideas of John Dalton about the nature of atoms and elements, allowed for the expression of chemical behavior in quantitative terms.

The work of Lavoisier and Dalton has been extended to explain how and why atoms bond as they do. This episode in the history of science is a good example of how important advances often involve relatively small modifications of prior knowledge, in this case, how an essentially descriptive science about materials and processes was slowly transformed into a much more systematic study of the nature of materials and how they interact.

The map is organized around three strands—*scientific understanding, history of development,* and *nature of science*. In middle school, the focus is on understanding basic ideas about chemical reactions. In high school, ideas about the historical episode and the nature of science are emphasized.

Maps in Chapter 4: THE PHYSICAL SETTING are particularly relevant to the scientific understanding strand on this map, especially **CONSERVATION OF MATTER, ATOMS AND MOLECULES**, and **CHEMICAL REACTIONS** in *Atlas 1*. The **SCIENTIFIC WORLD VIEW** map is also closely related.

## NOTES

As with other maps in Chapter 10, knowledge of the underlying scientific principles is useful in learning about how scientists developed their ideas about a particular topic, in this case, about chemical reactions. The *scientific understanding* strand presents the central benchmarks and includes some new and revised benchmarks. The middle-school benchmark, "Substances react chemically in characteristic ways..." is derived from *National Science Education Standards*, and another new 6-8 benchmark "The idea of atoms explains chemical reactions..." is from *Science for All Americans*. Benchmark 4D/M5 has been significantly revised.

The *history of development* strand follows two interconnected stories, one tracing the idea of atoms and the other tracing the idea of elements. Most of the benchmarks in this strand have been revised and several new ones have been added to make the presentation of this episode more accurate and coherent. For example, the idea of phlogiston is introduced in the new benchmark 10F/M7, and benchmark 10F/M4 has been revised to note how Lavoisier's experiments exposed insurmountable problems with the phlogiston theory.

The *nature of science* strand deals with different traditions in science and ways in which scientific views change over time.

## TIMELINE

**400 BCE**

440 BCE Leucippus of Miletus characterizes the world as consisting of atoms and a void.

420 BCE Democritus of Abdera develops Leucippus's atomic theory: Atoms vibrate in solids and exist in space.

**100 BCE**

100 BCE Hero of Alexandria explains that the four elements (earth, air, fire, and water) consist of atoms.

**1600**

1658 Henry More argues that atoms are the fundamental unit of matter and are too small to see.

1661 Robert Boyle gives the first precise definitions of a chemical element and chemical reaction in terms of discrete quantities, leading scientists to abandon the Greek idea of four elements.

**1700**

1730 Antoine Lavoisier and Pierre-Simon Laplace conclude that respiration is a form of combustion.

1786 Lavoisier publishes *Method of Chemical Nomenclature,* which describes his system of naming chemical compounds.

1789 Lavoisier publishes *Elementary Treatise of Chemistry*, which states the law of conservation of mass.

1799 Joseph Louis Proust describes the law of definite proportions.

**1800**

1805 John Dalton publishes a paper that describes the law of multiple proportions, suggests that all matter is made up of atoms, and provides a table of atomic weights.

1808 Joseph-Louis Gay-Lussac announces that gases combine chemically in definite proportions according to their volumes.

1811 Amedeo Avogadro proposes that equal volumes of different gases at the same temperature and pressure contain the same number of molecules.

1869 Dimitry Mendeleev and Julius Meyer formulate the periodic law.

**1900**

1916 Gilbert Lewis states that the chemical bond consists of two electrons held jointly by two atoms.

1931 Linus Pauling details rules of chemical bonding.

9-12

Since Lavoisier and Dalton, the system for describing chemical reactions has been vastly extended to account for the configuration taken by atoms when they bond to one another and to describe the inner workings of atoms that account for why they bond as they do. 10F/H5*

From time to time, major shifts occur in the scientific view of how things work. More often, however, the changes that take place in the body of scientific knowledge are small modifications of prior knowledge. Continuity and change are persistent features of science. 1A/H2*

Scientific Theories (v. 1, p. 21)
Scientific World View (p. 5)
Classical Mechanics (p. 75)
Relativity (p. 77)

4D/H1*

Dalton figured out how the relative weights of the atoms could be determined experimentally. His idea that every substance had a unique atomic composition provided an explanation for why substances were made up of elements in specific proportions. 10F/H4**

There are different traditions in science about what is investigated and how, but they all share a commitment to the use of logical arguments based on empirical evidence. 1B/H4*

Lavoisier pioneered a new approach to chemistry based on the modern definition of an element and quantitative methods. His system for naming substances and describing their reactions in terms of the elements that make them up contributed to the rapid growth of chemistry by enabling scientists everywhere to share their findings about chemical reactions without ambiguity. 10F/H2*

In the early 1800s, British chemist and physicist John Dalton united the concepts of atoms and elements. He proposed two ideas that laid the groundwork for modern chemistry: first, that elements are formed from small, indivisible particles called atoms, which are identical for a given element but different from any other element; and second, that chemical compounds are formed from atoms by combining a definite number of each type of atom to form one molecule of the compound. 10F/H3*

Scientific World View (p. 5)
Evidence and Reasoning in Inquiry (v. 1, p. 17)
Scientific Investigations (v. 1, p. 19)
Explaining Evolution (p. 85)

4D/M12**

1C/E2

In the late 1700s and early 1800s, the idea of atoms reemerged in response to questions about the structure of matter, the nature of fire, and the basis of chemical phenomena. 10F/H1**

6-8

Flow of Energy in Ecosystems (v. 1, p. 79)
Flow of Matter in Ecosystems (v. 1, p. 77)
Materials Science (p. 55)

Antoine Lavoisier's work was based on the idea that when materials react with each other, many changes can take place but that in every case the total amount of mass afterward is the same as before. He successfully tested the concept of conservation of mass by conducting a series of experiments in which he carefully measured all the substances involved in burning, including the gases used and those given off. As a result, the phlogiston theory was replaced by a theory based on the role of oxygen in burning. 10F/M4*

Scientific knowledge is subject to modification as new information challenges prevailing theories and as a new theory leads to looking at old observations in a new way. 1A/M2

The idea of atoms explains chemical reactions: When substances interact to form new substances, the atoms that make up the molecules of the original substances combine in new ways. 4D/M13**

Atoms may link together in well-defined molecules, or may be packed together in crystal patterns. Different arrangements of atoms into groups compose all substances and determine the characteristic properties of substances. 4D/M1cd*

4D/M7b

Constancy (p. 95)
Conservation of Matter (v. 1, p. 57)
Flow of Matter in Ecosystems (v. 1, p. 77)

Up until the 1700s, the prevailing idea was that materials contained a substance called phlogiston, which had mass, and that when an object burned, it gave off phlogiston that carried away the mass. This view confirmed what people saw: When a piece of wood was burned, all that was left was a residue of ashes that had far less mass than the wood. 10F/M7** (SFAA)

Atoms and Molecules (v. 1, p. 55)
Conservation of Matter (v. 1, p. 57)
States of Matter (v. 1, p. 59)

No matter how substances within a closed system interact with one another, or how they combine or break apart, the total mass of the system remains the same. 4D/M7a*

An important kind of reaction between substances involves the combination of oxygen with something else—as in burning or rusting. 4D/M6b*

Substances react chemically in characterstic ways with other substances to form new substances with different characteristic properties. 4D/M11** (NSES)

Chemical Reactions (v. 1, p. 61)

4D/M1b*

Flow of Energy in Ecosystems (v. 1, p. 79)

More than 2000 years ago, Greek philosophers formulated the idea of atoms as being simple particles beyond the reach of human senses and with the existence of a void between them. 10F/M1a*

Aristotle and other thinkers held that four elements (fire, air, water, and earth) composed all of matter. While the idea of atoms received little attention, Aristotle's four elements became the dominant idea. 10F/M1b*

Materials Science (p. 55)

All matter is made up of atoms, which are far too small to see directly through a microscope. 4D/M1a

Chemical elements are those substances that do not break down during normal laboratory reactions involving such treatments as heating, exposure to electric current, or reaction with acids. All substances from living and nonliving things can be broken down to a set of about 100 elements, but since most elements tend to combine with others, few elements are found in their pure form. 4D/M5*

Splitting the Atom (p. 83)
Atoms and Molecules (v. 1, p. 55)

Ancient civilizations in China, India, the Middle East, and the Mediterranean developed explanations about matter not by experimentation but by observing the natural world and by using logic and rational thought. 10F/M6**

scientific understanding    history of development    nature of science

# SPLITTING THE ATOM (10G)

At the beginning of the 19th century, it was accidentally discovered that a compound of uranium was able to darken a wrapped and unexposed photographic plate in the same way that light would darken the plate. Efforts by researchers Marie and Pierre Curie to explain this phenomenon led to a greater understanding of the structure of matter and the discovery of two new radioactive elements, polonium and radium. Later, Ernest Rutherford found that radioactive uranium spontaneously splits itself into particles of different weights and concluded that an atom must be made of different parts. Other scientists determined that when uranium is intentionally irradiated by neutrons, isotopes of various elements are produced. Finally, the splitting of the nucleus of an atom was found to release enormous amounts of energy. Scientists soon realized that the extra neutrons released by this splitting, or fission, could induce more fission, resulting in a chain reaction.

In addition to expanding our knowledge of the structure of matter, the study of radioactivity has also led to the development of nuclear weapons and nuclear reactors for generating power and to many other uses of radioactive materials. This historical episode illustrates vividly the idea that human inventiveness brings new risks as well as improvements to human existence.

The map is organized around three strands—*scientific understanding*, *history of development*, and *nature of science*. There is only one benchmark at the middle-school level; all others are introduced in high school.

Several maps in Chapter 4: THE PHYSICAL SETTING in *Atlas 1* have links to this map. In particular, the maps on **ATOMS AND MOLECULES** and **CONSERVATION OF MATTER** are quite relevant, as are maps in Chapter 1: THE NATURE OF SCIENCE published in both volumes of *Atlas*.

## NOTES

The *scientific understanding* strand has a number of relatively complicated benchmarks dealing with the structure of atoms. The benchmarks appear on this map to provide support for the historical ideas. Prerequisite knowledge for these scientific ideas can be found on other maps in Chapter 4, such as **ATOMS AND MOLECULES** in *Atlas 1*. Benchmarks 4D/H1, 4D/H2, and 4D/H3 in this strand have all been revised, and some ideas that were originally in benchmarks 4D/H2 and 4D/H3 have been moved to benchmark 4D/H1 “Atoms are made of a positively charged nucleus…” so that it fully describes the structure of the atom.

In the *history of development* strand, some of the ideas in benchmarks 10G/H1a and 10G/H1b were previously found in benchmarks at the 6-8 grade range but have been moved to benchmarks in the 9-12 range where they are more relevant to other ideas being addressed in high school. Benchmarks 10G/H2, 10G/H3a and 10G/H3b have all been revised to provide a more accurate and clear description of events.

Although the five benchmarks in the *nature of science* strand are not directly connected to each other, they can all contribute to the study of this historical episode. Most of the benchmarks in this strand touch on issues regarding the application of scientific knowledge and the relationships between science, technology, and society.

## TIMELINE

**1800**

1886 Henri Becquerel discovers radioactivity in uranium.

1895 Wilhelm Röntgen discovers X-rays.

1898 Marie and Pierre Curie discover and isolate radium and polonium.

1899 Becquerel shows that radioactivity consists of charged particles.

**1900**

1901 Ernest Rutherford publishes a paper describing elements that emit particles and form other elements.

1911 Rutherford discovers that atoms have a small, dense nucleus.

1913 Frederick Soddy discovers different forms of the same element, now known as isotopes.

1918 Rutherford discovers the proton.

1920 Rutherford suggests the existence of the neutron to keep an atom’s nucleus stable and to account for the existence of isotopes.

1934 Frederic and Irene Joliot-Curie produce radiation artificially by bombarding certain nuclei with alpha particles.

1934 Enrico Fermi and others show that nuclear reactions can be controlled.

1938 Otto Hahn and Fritz Strassmann discover that uranium can be converted into barium; Lise Meitner determines that splitting of the nuclei of uranium atoms is the cause.

1939 Albert Einstein writes President Franklin D. Roosevelt to encourage U.S. research on the possible use of nuclear fission in a weapon.

1942 The first sustained chain reaction is produced at the University of Chicago under Fermi’s supervision.

1945 The United States drops atomic bombs on Hiroshima and Nagasaki, prompting Japan to surrender and ending World War II in Asia.

1946 Willard Libby develops radiocarbon dating.

1952 The United States conducts the first test of a fusion explosion in the Marshall Islands.

1955 Emilio Segré discovers anti-protons.

1956 The first commercial nuclear power station is opened in England.

1986 An accident at the Chernobyl nuclear power plant exposes millions of people to radiation.

9-12

Human inventiveness has brought new risks as well as improvements to human existence. 3C/H5

The Industrial Revolution (p. 89)
Public Perception of Science (p. 105)

Nuclear weapons and energy remain matters of public concern and controversy. 10G/H4b

When applications of research could pose risks to society, scientists' decisions to participate in that research are based on personal as well as professional ethics. 1C/H5b*

Some scientists and engineers are comfortable working in situations in which some secrecy is required, but others prefer not to do so. It is generally regarded as a matter of individual choice and ethics, not one of professional ethics. 3C/H2*

Radioactivity has many uses other than generating energy, including in medicine, industry, and scientific research in many different fields. 10G/H5

3A/H1*

8F/H9**

Science and Society (p. 9)

A massive effort went into developing the technology for the production of nuclear fission bombs used against Japan in World War II. The hydrogen bomb which uses nuclear fusion was developed shortly after World War II. Another important development of this era was the nuclear reactor, in which nuclear energies are released in a controlled fashion for the production of electrical energy. 10G/H4a*

Funding influences the direction of science by virtue of the decisions that are made on which research to support. Research funding comes from various federal government agencies, industry, and private foundations. 1C/H8

Energy Resources (p. 59)

Nuclear reactions release energy without the combustion products of burning fuels, but the radioactivity of fuels and their by-products poses other risks. 8C/H3*

The Scientific Community (p. 7)

4G/H6*

Relativity (p. 77)

Enrico Fermi, an Italian working with colleagues in the United States, showed that the extra neutrons trigger more fissions in uranium nuclei and so create a sustained chain reaction in which a prodigious amount of energy is given off. 10G/H3c*

Energy Transformations (p. 25)

Energy is released whenever the nuclei of very heavy atoms, such as uranium or plutonium, split into middleweight ones, or when very light nuclei, such as those of hydrogen and helium, combine into heavier ones. For a given quantity of a substance, the energy released in a nuclear reaction is very much greater than the energy given off in a chemical reaction. 4E/H6*

In proposing that a uranium nucleus could split, Meitner had to account for the energy that would be released when the newly formed nuclei accelerated away from each other. She suggested it could come from the conversion of mass to energy predicted by Einstein's theory of special relativity. 10G/H3b*

10C/H3*

Otto Hahn and Fritz Strassmann, German chemists, found that when they struck uranium with a beam of neutrons, barium was unexpectedly produced. Austrian physicist Lise Meitner, who had earlier worked with Hahn and Strassmann on the decay patterns of uranium, was the first to suggest that the barium could have resulted from the splitting of the uranium nucleus into two middleweight nuclei and one or two extra neutrons. 10G/H3a*

Science disciplines differ from one another in what is studied, techniques used, and outcomes sought, but they share a common purpose and philosophy, and all are part of the same scientific enterprise. Although each discipline provides a conceptual structure for organizing and pursuing knowledge, many problems are studied by scientists using information and skills from many disciplines. Disciplines do not have fixed boundaries, and it happens that new scientific disciplines are being formed where existing ones meet and that some subdisciplines spin off to become new disciplines in their own right. 1C/H4

The nucleus of radioactive isotopes is unstable and spontaneously decays, emitting particles and/or wavelike radiation. It cannot be predicted exactly when, if ever, an unstable nucleus will decay, but a large group of identical nuclei decay at a predictable rate. This predictability of decay rate allows radioactivity to be used for estimating the age of materials that contain radioactive substances. 4D/H4

Ernest Rutherford of New Zealand and his colleagues discovered that the heavy radioactive element uranium spontaneously splits itself into a slightly lighter nucleus and a very light helium nucleus, leading to the realization that one kind of atom may change into another kind, and that therefore an atom must be made up of smaller parts. 10G/H2*

Atoms and Molecules (v. 1, p. 55)

The number of protons in the nucleus determines what an atom's electron configuration can be and so defines the element. An atom's electron configuration, particularly the outermost electrons, determines how the atom can interact with other atoms. Atoms form bonds to other atoms by transferring or sharing electrons. 4D/H2*

Although neutrons have little effect on how an atom interacts with other atoms, the number of neutrons does affect the mass and stability of the nucleus. Isotopes of the same element have the same number of protons (and therefore of electrons) but differ in the number of neutrons. 4D/H3*

In their laboratory in France, Marie Curie and her husband, Pierre Curie, isolated two new elements that caused most of the radioactivity of the uranium mineral. They named one radium because it gave off powerful, invisible rays, and the other polonium in honor of Madame Curie's country of birth, Poland. 10G/H1b*

Atoms and Molecules (v. 1, p. 55)
Electricity and Magnetism (p. 27)

Atoms are made of a positively charged nucleus surrounded by negatively charged electrons. The nucleus is a tiny fraction of the volume of an atom but makes up almost all of its mass. The nucleus is composed of protons and neutrons which have roughly the same mass but differ in that protons are positively charged while neutrons have no electric charge. 4D/H1*

The unexpected discovery that minerals containing uranium darken photographic film, as light does, led to the idea of radioactivity. 10G/H1a*

1B/H7

6-8

Chemical Reactions (v. 1, p. 61)

Chemical elements are those substances that do not break down during normal laboratory reactions involving such treatments as heating, exposure to electric current, or reaction with acids. All substances from living and nonliving things can be broken down to a set of about 100 elements, but since most elements tend to combine with others, few elements are found in their pure form. 4D/M5*

The Chemical Revolution (p. 81)
Atoms and Molecules (v. 1, p. 55)

scientific understanding | history of development | nature of science

# EXPLAINING EVOLUTION (10H)

The dramatic effect of Darwin's *On the Origin of Species* on biology can be traced to several factors: The argument Darwin presented was sweeping, yet clear and understandable; his line of argument was supported with a wealth of biological and fossil evidence; his comparison of natural selection to the "artificial selection" used in animal breeding was persuasive; and the argument provided a unifying framework for guiding future research.

There are some people who reject the concept of evolution—not on scientific grounds but on the basis of what they take to be its unacceptable implications: that human beings and other species have common ancestors and are therefore related; that humans and other organisms might have resulted from a process that lacks direction and purpose; and that human beings, like other animals, are engaged in a struggle for survival and reproduction. And for some people, the concept of evolution violates the biblical account of the special (and separate) creation of humans and all other species.

The study of this history provides a good opportunity to emphasize the importance in science of careful observation and description and to illustrate that not all scientific advances depend on experimentation. The history of this episode should also take note of people's acceptance and rejection of ideas about evolution, in the past and in the present.

The map is organized around four strands—*scientific ideas about evolution, development and acceptance of scientific ideas about evolution, determining the age of the earth,* and *nature of science*. Two benchmarks in the elementary- and middle-school grades provide a foundation for understanding natural selection. But because of the complexity of the scientific evidence and arguments that must be examined, a thorough understanding of species evolution probably cannot be achieved earlier than high school.

Maps on **BIOLOGICAL EVOLUTION** and **NATURAL SELECTION** in *Atlas 1* are extremely relevant to this map. Other closely linked maps are in Chapter 1: THE NATURE OF SCIENCE in *Atlas 1* and in this volume.

## NOTES

The *scientific ideas about evolution* strand presents just a few of the ideas needed to understand the basic mechanisms that underpin modern theory on evolution. More complete presentations of the learning progressions involving these ideas can be found on the **DIVERSITY OF LIFE** map in this volume and on the **NATURAL SELECTION** and **BIOLOGICAL EVOLUTION** maps in *Atlas 1*. The 3-5 benchmark "There are millions of different species..." is a new statement based on text from *Science for All Americans*, and the 6-8 benchmark "Most species that have lived..." is a new statement based on *National Science Education Standards*. Together, the two statements describe important prerequisite ideas that contribute to an understanding of the processes of natural selection and how Darwin discovered them.

In the *development and acceptance of scientific ideas about evolution* strand, the benchmarks have been revised to improve their clarity and accuracy.

Ideas in the *determining the age of the earth* strand are important because they explain how Darwin's arguments depended on the notion that the earth was much older than was commonly believed and that the types of processes that occurred in the past are the same as those that take place today.

Several of the benchmarks in the *nature of science* strand can also be found on **THE COPERNICAN REVOLUTION** map. Both maps focus on discoveries made through the study of phenomena as they occurred in nature rather than during controlled experiments in the laboratory. The ideas of Copernicus and Darwin challenged people's fundamental understanding of how the world works. Nevertheless, the scientific community readily accepted the new ideas because of their explanatory and predictive powers.

## TIMELINE

**1700**

1744 Georges Buffon states that the earth is at least 75,000 years old.

1745 Pierre Maupertius proposes the idea of common descent.

1755 Buffon suggests that species are shaped by their environment and that these changes are heritable.

**1800**

1809 Jean Lamarck is the first important proponent of evolution, though his theory suggests that characteristics acquired during an individual's lifetime can be passed on.

1833 Charles Lyell publishes the third, and final, volume of *Principles of Geology*.

1835 Charles Darwin completes his four-year voyage on the HMS Beagle.

1859 Darwin publishes *On the Origin of Species by Means of Natural Selection*.

1866 Gregor Mendel interprets heredity in terms of discrete particles.

1871 Darwin publishes *The Descent of Man*, suggesting that the evolution of humans and animals differs only in degree, not in kind.

1883 August Weismann refutes Lamarck's theory that acquired characteristics can be inherited, arguing instead that information flows from sex cells to body cells, not the reverse.

**1900**

1900 The significance of Mendel's work is recognized in three independent accounts.

1902 William Bateson demonstrates that Mendel's principles, derived from studying plants, also apply to animals.

1904 Thomas Nuttall infers, based on similarities in their proteins, that humans are more closely related to old-world than to new-world primates, which corresponds to Darwin's conclusions based on structural similarities.

1944 Oswald Avery, Colin MacLeod, Maclyn McCarty, and Frederick Griffith establish that DNA is the genetic material responsible for heredity.

1953 James Watson and Francis Crick work out the structure of the DNA double helix and show how a molecule could carry functional genetic information.

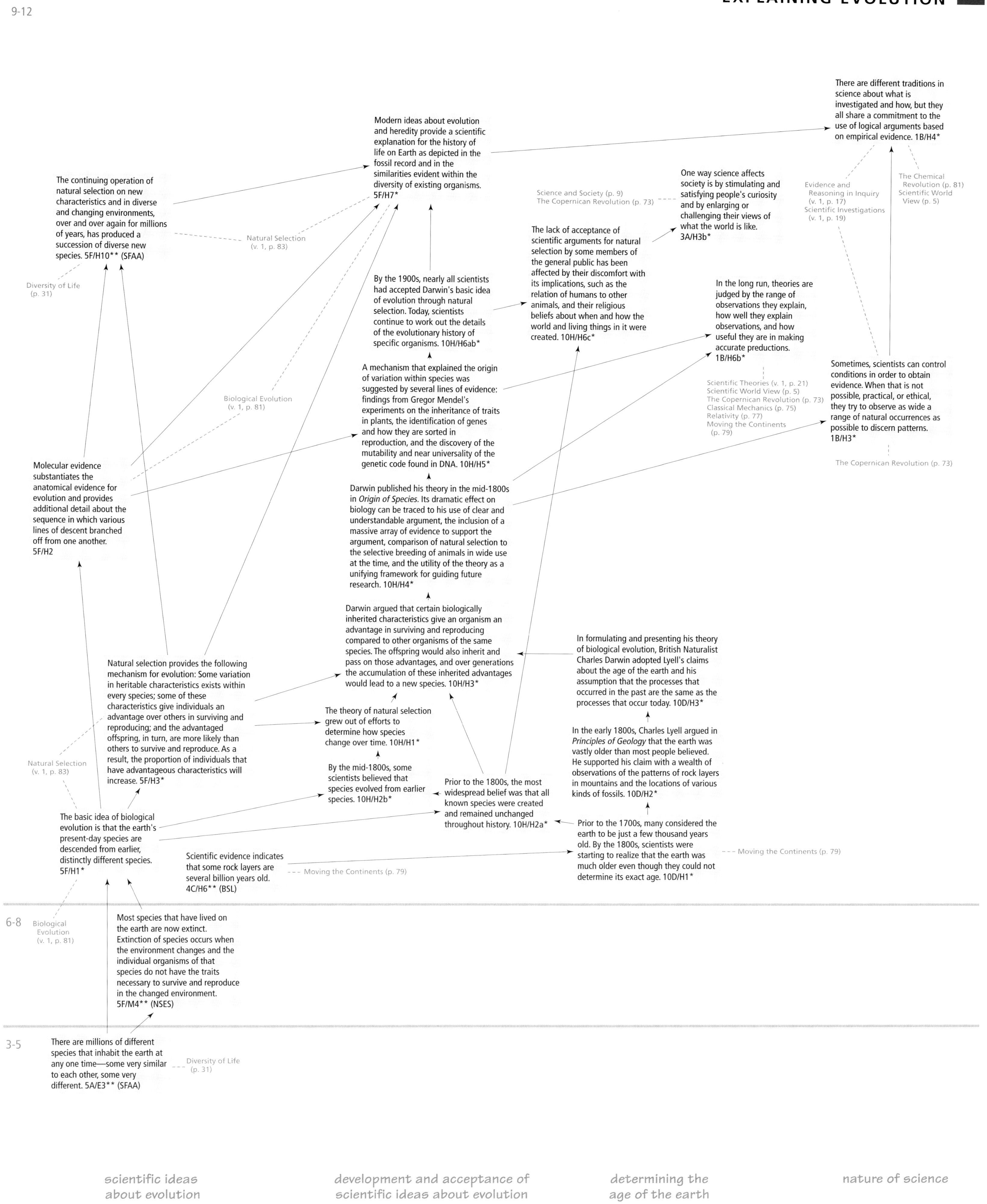
9-12
There are different traditions in science about what is investigated and how, but they all share a commitment to the use of logical arguments based on empirical evidence. 1B/H4*
Modern ideas about evolution and heredity provide a scientific explanation for the history of life on Earth as depicted in the fossil record and in the similarities evident within the diversity of existing organisms. 5F/H7*
The continuing operation of natural selection on new characteristics and in diverse and changing environments, over and over again for millions of years, has produced a succession of diverse new species. 5F/H10** (SFAA)
Natural Selection (v. 1, p. 83)
Diversity of Life (p. 31)
One way science affects society is by stimulating and satisfying people's curiosity and by enlarging or challenging their views of what the world is like. 3A/H3b*
Science and Society (p. 9)
The Copernican Revolution (p. 73)
Evidence and Reasoning in Inquiry (v. 1, p. 17)
Scientific Investigations (v. 1, p. 19)
The Chemical Revolution (p. 81)
Scientific World View (p. 5)
The lack of acceptance of scientific arguments for natural selection by some members of the general public has been affected by their discomfort with its implications, such as the relation of humans to other animals, and their religious beliefs about when and how the world and living things in it were created. 10H/H6c*
By the 1900s, nearly all scientists had accepted Darwin's basic idea of evolution through natural selection. Today, scientists continue to work out the details of the evolutionary history of specific organisms. 10H/H6ab*
In the long run, theories are judged by the range of observations they explain, how well they explain observations, and how useful they are in making accurate preductions. 1B/H6b*
Scientific Theories (v. 1, p. 21)
Scientific World View (p. 5)
The Copernican Revolution (p. 73)
Classical Mechanics (p. 75)
Relativity (p. 77)
Moving the Continents (p. 79)
Sometimes, scientists can control conditions in order to obtain evidence. When that is not possible, practical, or ethical, they try to observe as wide a range of natural occurrences as possible to discern patterns. 1B/H3*
The Copernican Revolution (p. 73)
A mechanism that explained the origin of variation within species was suggested by several lines of evidence: findings from Gregor Mendel's experiments on the inheritance of traits in plants, the identification of genes and how they are sorted in reproduction, and the discovery of the mutability and near universality of the genetic code found in DNA. 10H/H5*
Biological Evolution (v. 1, p. 81)
Molecular evidence substantiates the anatomical evidence for evolution and provides additional detail about the sequence in which various lines of descent branched off from one another. 5F/H2
Darwin published his theory in the mid-1800s in Origin of Species. Its dramatic effect on biology can be traced to his use of clear and understandable argument, the inclusion of a massive array of evidence to support the argument, comparison of natural selection to the selective breeding of animals in wide use at the time, and the utility of the theory as a unifying framework for guiding future research. 10H/H4*
Darwin argued that certain biologically inherited characteristics give an organism an advantage in surviving and reproducing compared to other organisms of the same species. The offspring would also inherit and pass on those advantages, and over generations the accumulation of these inherited advantages would lead to a new species. 10H/H3*
In formulating and presenting his theory of biological evolution, British Naturalist Charles Darwin adopted Lyell's claims about the age of the earth and his assumption that the processes that occurred in the past are the same as the processes that occur today. 10D/H3*
Natural selection provides the following mechanism for evolution: Some variation in heritable characteristics exists within every species; some of these characteristics give individuals an advantage over others in surviving and reproducing; and the advantaged offspring, in turn, are more likely than others to survive and reproduce. As a result, the proportion of individuals that have advantageous characteristics will increase. 5F/H3*
The theory of natural selection grew out of efforts to determine how species change over time. 10H/H1*
In the early 1800s, Charles Lyell argued in Principles of Geology that the earth was vastly older than most people believed. He supported his claim with a wealth of observations of the patterns of rock layers in mountains and the locations of various kinds of fossils. 10D/H2*
Natural Selection (v. 1, p. 83)
By the mid-1800s, some scientists believed that species evolved from earlier species. 10H/H2b*
Prior to the 1800s, the most widespread belief was that all known species were created and remained unchanged throughout history. 10H/H2a*
The basic idea of biological evolution is that the earth's present-day species are descended from earlier, distinctly different species. 5F/H1*
Prior to the 1700s, many considered the earth to be just a few thousand years old. By the 1800s, scientists were starting to realize that the earth was much older even though they could not determine its exact age. 10D/H1*
Moving the Continents (p. 79)
Scientific evidence indicates that some rock layers are several billion years old. 4C/H6** (BSL)
Moving the Continents (p. 79)
6-8
Biological Evolution (v. 1, p. 81)
Most species that have lived on the earth are now extinct. Extinction of species occurs when the environment changes and the individual organisms of that species do not have the traits necessary to survive and reproduce in the changed environment. 5F/M4** (NSES)
3-5
There are millions of different species that inhabit the earth at any one time—some very similar to each other, some very different. 5A/E3** (SFAA)
Diversity of Life (p. 31)
scientific ideas about evolution
development and acceptance of scientific ideas about evolution
determining the age of the earth
nature of science

# HISTORICAL PERSPECTIVES

## DISCOVERING GERMS (10I)

During the 19th century, evidence started to accumulate that microorganisms might be responsible for causing some diseases. Even though scientists had been aware of the existence of microscopically small plants and animals since the development of the microscope in the 1600s, microorganisms were not immediately linked to disease because most do not cause disease and many are beneficial. Louis Pasteur demonstrated that spoilage and fermentation occur with exposure to microorganisms in the air and that spoilage could be avoided by keeping organisms out or by killing them. He also showed that infection by disease organisms, including intentional inoculation with disease organisms, caused the body to build up immunity to that disease. However, the actual linking of a specific, identifiable microorganism to a particular disease had to await the work of scientists such as Robert Koch, who isolated a bacterium linked to anthrax. Using Koch's rigorous set of criteria for determining which specific microorganism is responsible for a particular disease, scientists identified many more disease organisms in the ensuing years.

The discovery of disease-producing microorganisms has had an enormous effect on current health practices, including the safe handling of food and water, the pasteurization of milk, and the use of antiseptic surgical techniques and antibiotics. The discoveries of Pasteur and others came from efforts to solve very practical problems involving food spoilage and plant diseases. This episode in the history of science is a good example of how the desire to solve practical problems can be a powerful incentive in the development of scientific knowledge.

The map is organized around two strands—*development of germ theory* and *nature of science*. Most of the benchmarks on the map are concentrated in middle school, with only two benchmarks at the elementary-school level and two at the high-school level. The K-2 and 3-5 benchmarks introduce ideas about germs, disease, and immunity and are necessary prerequisites for understanding the discovery of germs.

In Chapter 8: THE DESIGNED WORLD, the map on **HEALTH TECHNOLOGY** is closely related to this map, and the two maps share several benchmarks. Maps in Chapter 1: THE NATURE OF SCIENCE in *Atlas 1* and this volume are also relevant.

### NOTES

Several of the benchmarks in the *development of germ theory* strand have been revised to make the description of this historical episode more accurate. For example, Koch's contributions to understanding the role of germs in disease are now explicitly mentioned in benchmarks 10I/M2d and 10I/M4. Text from *Science for All Americans* was used to craft the new 6-8 benchmark "The improvement of microscope lenses..." to explain why the link between microorganisms and disease was not obvious. Another new 6-8 benchmark "After the discovery of germs..." was derived from *Science for All Americans* and describes the ongoing study of the many different kinds of microorganisms.

The *nature of science* strand is made up of two separate threads depicting different aspects of scientific discovery and its impact on society.

### TIMELINE

**1600**

1650s Thomas Sydenham proposes that diseases result from organisms inside a host and argues that direct observation and classification should be the basis for determining the nature of disease.

1676 Anton van Leeuwenhoek observes microorganisms through a microscope.

**1700**

1765 Lazzaro Spallanzani establishes that microbes arise only from existing organisms (but spontaneous generation continues to find adherents until Louis Pasteur's discoveries a century later).

1796 Edward Jenner confirms that exposure to cowpox renders milkmaids immune to smallpox.

**1800**

1853 Florence Nightingale dramatically reduces the death rate in hospitals through a regimen of cleanliness.

1861 Ignaz Semmelweis deduces that childbirth fever is transmitted on the hands of doctors during their examinations of patients.

1862 Pasteur develops a process for heating liquids to kill bacteria in them, a process now referred to as pasteurization.

1876 Robert Koch develops a bacterial-staining technique and uses it to identify the bacterium that causes anthrax and, later, several other disease-causing bacteria to confirm the germ theory of disease.

1879 Pasteur accidentally discovers a method to weaken bacteria so that they fail to cause a disease but still make the infected organism immune to future infections.

1881 Pasteur conducts a public demonstration of a method for vaccinating against anthrax.

1882 Dimitri Ivanovsky demonstrates that tobacco mosaic disease is caused by an agent (now known as a virus) that self-replicates like bacteria but is far smaller.

1882 Koch identifies the bacterium that causes tuberculosis.

**1900**

1928 Alexander Fleming discovers the antibiotic penicillin.

1954 Jonas Salk develops a killed virus vaccine against polio.

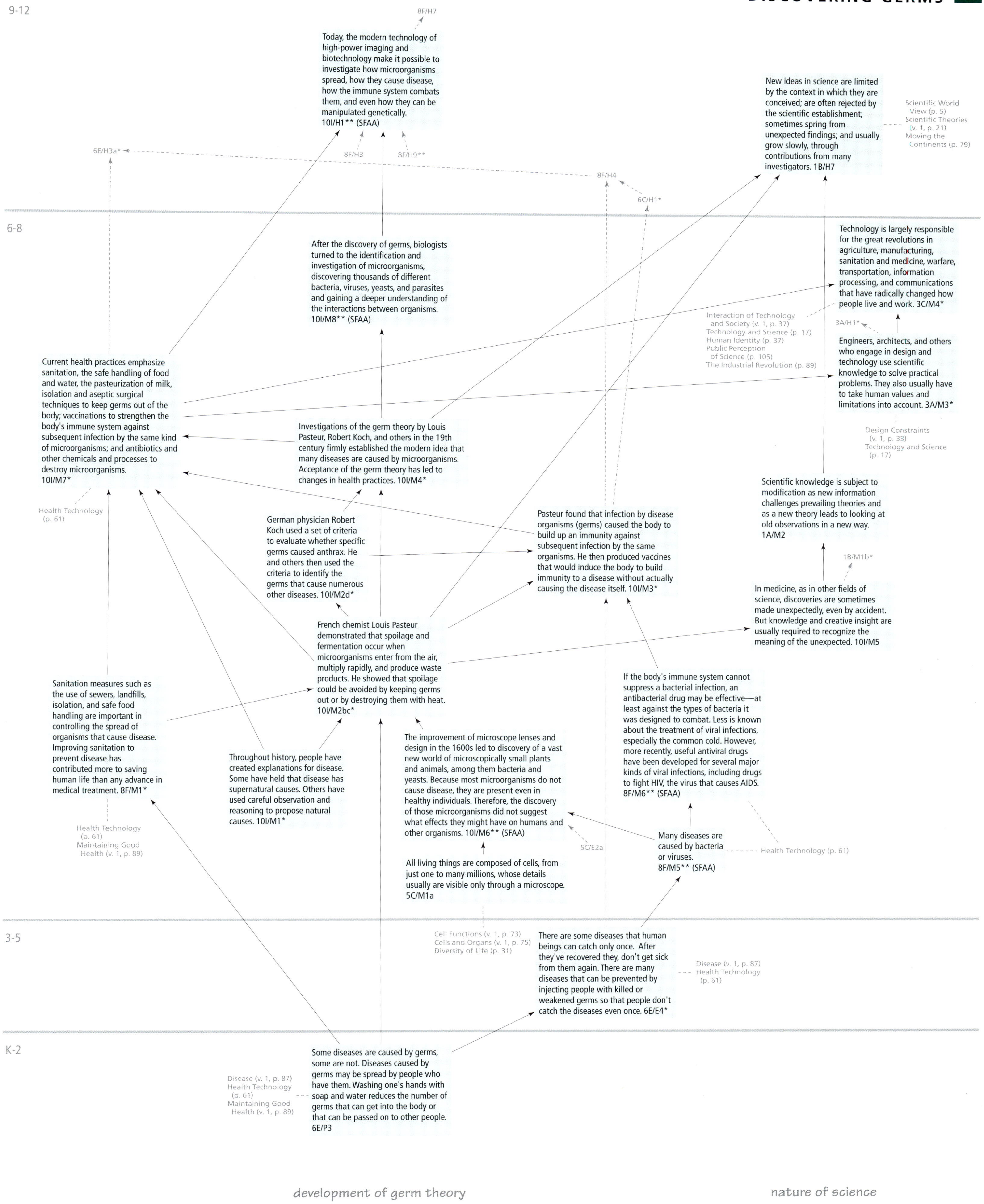
9-12
8F/H7
Today, the modern technology of high-power imaging and biotechnology make it possible to investigate how microorganisms spread, how they cause disease, how the immune system combats them, and even how they can be manipulated genetically. 10I/H1** (SFAA)
8F/H3
8F/H9**
6E/H3a*
8F/H4
6C/H1*
New ideas in science are limited by the context in which they are conceived; are often rejected by the scientific establishment; sometimes spring from unexpected findings; and usually grow slowly, through contributions from many investigators. 1B/H7
Scientific World View (p. 5)
Scientific Theories (v. 1, p. 21)
Moving the Continents (p. 79)
6-8
After the discovery of germs, biologists turned to the identification and investigation of microorganisms, discovering thousands of different bacteria, viruses, yeasts, and parasites and gaining a deeper understanding of the interactions between organisms. 10I/M8** (SFAA)
Technology is largely responsible for the great revolutions in agriculture, manufacturing, sanitation and medicine, warfare, transportation, information processing, and communications that have radically changed how people live and work. 3C/M4*
Interaction of Technology and Society (v. 1, p. 37)
Technology and Science (p. 17)
Human Identity (p. 37)
Public Perception of Science (p. 105)
The Industrial Revolution (p. 89)
3A/H1*
Engineers, architects, and others who engage in design and technology use scientific knowledge to solve practical problems. They also usually have to take human values and limitations into account. 3A/M3*
Design Constraints (v. 1, p. 33)
Technology and Science (p. 17)
Current health practices emphasize sanitation, the safe handling of food and water, the pasteurization of milk, isolation and aseptic surgical techniques to keep germs out of the body; vaccinations to strengthen the body's immune system against subsequent infection by the same kind of microorganisms; and antibiotics and other chemicals and processes to destroy microorganisms. 10I/M7*
Health Technology (p. 61)
Investigations of the germ theory by Louis Pasteur, Robert Koch, and others in the 19th century firmly established the modern idea that many diseases are caused by microorganisms. Acceptance of the germ theory has led to changes in health practices. 10I/M4*
Scientific knowledge is subject to modification as new information challenges prevailing theories and as a new theory leads to looking at old observations in a new way. 1A/M2
German physician Robert Koch used a set of criteria to evaluate whether specific germs caused anthrax. He and others then used the criteria to identify the germs that cause numerous other diseases. 10I/M2d*
Pasteur found that infection by disease organisms (germs) caused the body to build up an immunity against subsequent infection by the same organisms. He then produced vaccines that would induce the body to build immunity to a disease without actually causing the disease itself. 10I/M3*
1B/M1b*
In medicine, as in other fields of science, discoveries are sometimes made unexpectedly, even by accident. But knowledge and creative insight are usually required to recognize the meaning of the unexpected. 10I/M5
French chemist Louis Pasteur demonstrated that spoilage and fermentation occur when microorganisms enter from the air, multiply rapidly, and produce waste products. He showed that spoilage could be avoided by keeping germs out or by destroying them with heat. 10I/M2bc*
Sanitation measures such as the use of sewers, landfills, isolation, and safe food handling are important in controlling the spread of organisms that cause disease. Improving sanitation to prevent disease has contributed more to saving human life than any advance in medical treatment. 8F/M1*
If the body's immune system cannot suppress a bacterial infection, an antibacterial drug may be effective—at least against the types of bacteria it was designed to combat. Less is known about the treatment of viral infections, especially the common cold. However, more recently, useful antiviral drugs have been developed for several major kinds of viral infections, including drugs to fight HIV, the virus that causes AIDS. 8F/M6** (SFAA)
Throughout history, people have created explanations for disease. Some have held that disease has supernatural causes. Others have used careful observation and reasoning to propose natural causes. 10I/M1*
The improvement of microscope lenses and design in the 1600s led to discovery of a vast new world of microscopically small plants and animals, among them bacteria and yeasts. Because most microorganisms do not cause disease, they are present even in healthy individuals. Therefore, the discovery of those microorganisms did not suggest what effects they might have on humans and other organisms. 10I/M6** (SFAA)
Health Technology (p. 61)
Maintaining Good Health (v. 1, p. 89)
5C/E2a
Many diseases are caused by bacteria or viruses. 8F/M5** (SFAA)
Health Technology (p. 61)
All living things are composed of cells, from just one to many millions, whose details usually are visible only through a microscope. 5C/M1a
3-5
Cell Functions (v. 1, p. 73)
Cells and Organs (v. 1, p. 75)
Diversity of Life (p. 31)
There are some diseases that human beings can catch only once. After they've recovered they, don't get sick from them again. There are many diseases that can be prevented by injecting people with killed or weakened germs so that people don't catch the diseases even once. 6E/E4*
Disease (v. 1, p. 87)
Health Technology (p. 61)
K-2
Disease (v. 1, p. 87)
Health Technology (p. 61)
Maintaining Good Health (v. 1, p. 89)
Some diseases are caused by germs, some are not. Diseases caused by germs may be spread by people who have them. Washing one's hands with soap and water reduces the number of germs that can get into the body or that can be passed on to other people. 6E/P3
development of germ theory
nature of science

# THE INDUSTRIAL REVOLUTION (10J)

The Industrial Revolution involved a radical shift in society from a rural handicraft economy to an urban manufacturing economy. At the heart of the Industrial Revolution was the invention and improvement of the steam engine, which transformed chemical energy into motion energy. This vastly increased the productivity of each worker and the amount of work that could be accomplished, but it also required a change in the structure of society as it concentrated labor in factories in urban centers. In addition to producing tremendous economic benefits for some, the dislocations brought about by the Industrial Revolution also often resulted in unhealthy working conditions, an erosion of the rural craft tradition, and an exploitation of child labor.

The changes that took place during the Industrial Revolution point to the development of new technology as a powerful catalyst for social transformation. In this case, the development of the steam engine, the machinery used in the factories, and various new modes of transportation radically changed how people live and work.

The map is organized around three strands—*technological understanding, history of development,* and *nature of technology*. In elementary school, the focus is on differences between craft- and manufacturing-based economies. In middle school, the emphasis shifts to ideas about particular technologies and practices that characterize industrial societies. By high school, the scope has broadened to include social and cultural changes brought about by the Industrial Revolution and impacts of technology on society in general.

The **MANUFACTURING** map is closely linked to many of the concepts on this map. Several maps from Chapter 3: THE NATURE OF TECHNOLOGY in both *Atlas 1* and this volume are also relevant.

## NOTES

Concepts described in the *technological understanding* strand are key to understanding the significance of the Industrial Revolution. All three of the benchmarks in this strand also appear on the **MANUFACTURING** map.

In the *history of development* strand, the 3-5 benchmark "In the 1700s, most manufacturing..." was drawn from a 6-8 benchmark to connect with other elementary-level ideas, including one about manufacturing described in benchmark 8B/E4. Other benchmarks in this strand have been edited to improve their accuracy and clarity.

The *nature of technology* strand includes benchmarks that address the interaction of technology and society.

## TIMELINE

**1600**

1698 Thomas Savery patents an engine that can be used to raise water from a mine by condensing steam to produce a vacuum.

**1700**

1709 Abraham Darby develops a coke-fired blast furnace that can produce iron more cost-effectively than coal-burning furnaces can.

1712 Thomas Newcomen demonstrates the first practical steam engine for industrial purposes.

1721 A large water-powered silk mill, which some consider to be the first factory, is built in Derby, England.

1733 John Kay patents the flying shuttle loom.

1762 Mathew Boulton establishes the Soho Manufactory near Birmingham, England.

1769 James Watt patents improvements to the Newcomen-style steam engine that makes it far more efficient.

1769 Nicolas-Joseph Cugnot demonstrates the first self-propelled steam vehicle.

1771 Richard Arkwright builds the first water-powered cotton mill in Cromford, England.

1775 Boulton and Watt form a partnership to produce steam engines at the Soho Foundry near Boulton's Soho Manufactory.

1785 Edmund Cartwright builds the first power loom.

1788 William Symington builds the first practical steamboat.

1794 Eli Whitney patents the cotton gin.

**1800**

1802 Richard Trevithick develops the first working steam locomotive.

1807 Robert Fulton builds a commercially viable paddle-wheel steamboat.

1812 Matthew Murray builds the Salamanca, the first commercially successful steam locomotive.

**1900**

1908 Henry Ford uses assembly-line manufacturing to produce the Model T automobile.

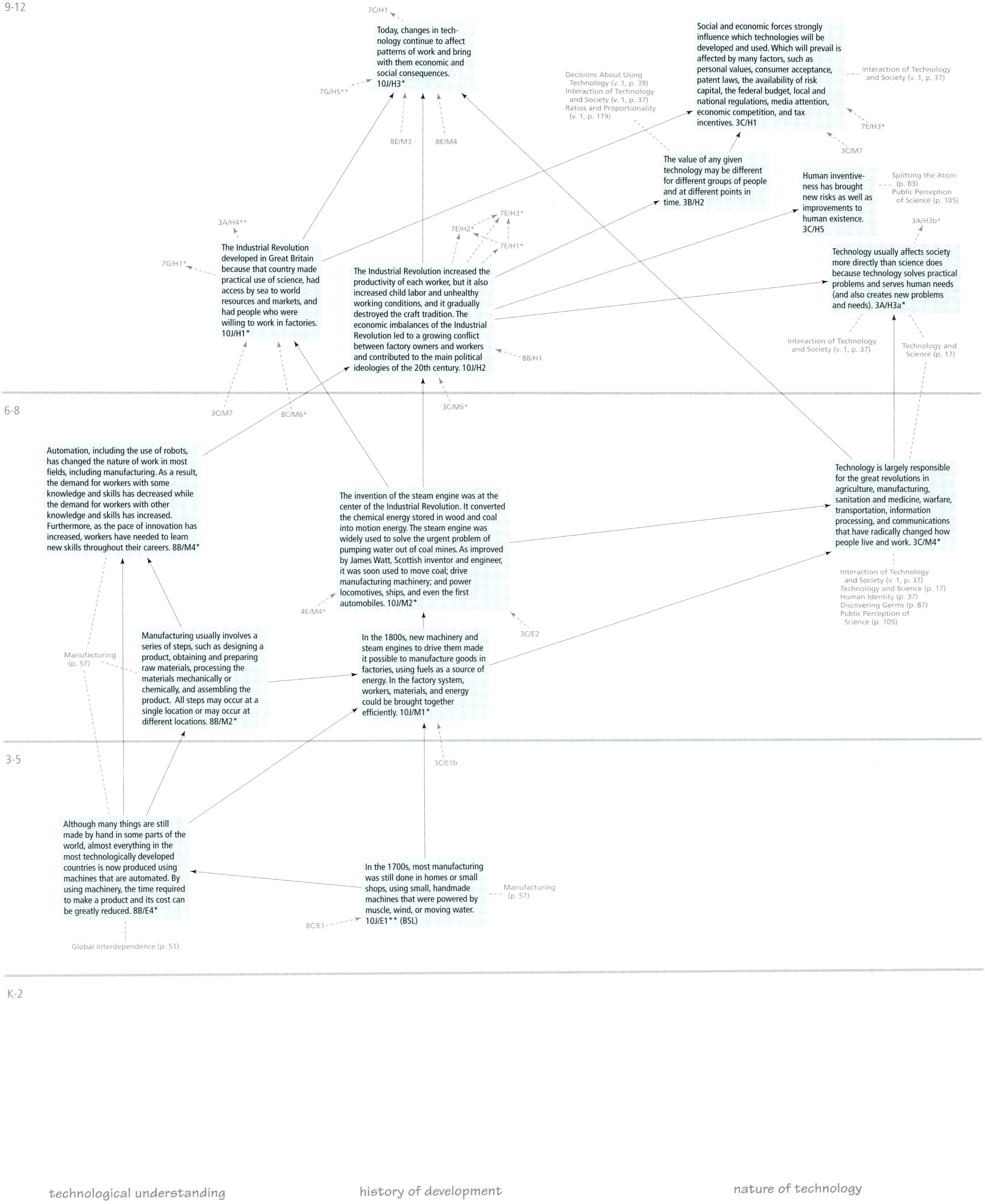
9-12
7C/H1
Today, changes in technology continue to affect patterns of work and bring with them economic and social consequences. 10J/H3*
7G/H5**
8E/M3
8E/M4
Decisions About Using Technology (v. 1, p. 39)
Interaction of Technology and Society (v. 1, p. 37)
Ratios and Proportionality (v. 1, p. 119)
Social and economic forces strongly influence which technologies will be developed and used. Which will prevail is affected by many factors, such as personal values, consumer acceptance, patent laws, the availability of risk capital, the federal budget, local and national regulations, media attention, economic competition, and tax incentives. 3C/H1
Interaction of Technology and Society (v. 1, p. 37)
7E/H3*
3C/M7
The value of any given technology may be different for different groups of people and at different points in time. 3B/H2
Human inventiveness has brought new risks as well as improvements to human existence. 3C/H5
Splitting the Atom (p. 83)
Public Perception of Science (p. 105)
3A/H3b*
3A/H4**
7E/H3*
7E/H2*
7E/H1*
7G/H1*
The Industrial Revolution developed in Great Britain because that country made practical use of science, had access by sea to world resources and markets, and had people who were willing to work in factories. 10J/H1*
The Industrial Revolution increased the productivity of each worker, but it also increased child labor and unhealthy working conditions, and it gradually destroyed the craft tradition. The economic imbalances of the Industrial Revolution led to a growing conflict between factory owners and workers and contributed to the main political ideologies of the 20th century. 10J/H2
8B/H1
Technology usually affects society more directly than science does because technology solves practical problems and serves human needs (and also creates new problems and needs). 3A/H3a*
Interaction of Technology and Society (v. 1, p. 37)
Technology and Science (p. 17)
3C/M7
8C/M6*
3C/M6*
6-8
Automation, including the use of robots, has changed the nature of work in most fields, including manufacturing. As a result, the demand for workers with some knowledge and skills has decreased while the demand for workers with other knowledge and skills has increased. Furthermore, as the pace of innovation has increased, workers have needed to learn new skills throughout their careers. 8B/M4*
The invention of the steam engine was at the center of the Industrial Revolution. It converted the chemical energy stored in wood and coal into motion energy. The steam engine was widely used to solve the urgent problem of pumping water out of coal mines. As improved by James Watt, Scottish inventor and engineer, it was soon used to move coal; drive manufacturing machinery; and power locomotives, ships, and even the first automobiles. 10J/M2*
4E/M4*
Technology is largely responsible for the great revolutions in agriculture, manufacturing, sanitation and medicine, warfare, transportation, information processing, and communications that have radically changed how people live and work. 3C/M4*
Interaction of Technology and Society (v. 1, p. 37)
Technology and Science (p. 17)
Human Identity (p. 37)
Discovering Germs (p. 87)
Public Perception of Science (p. 105)
3C/E2
Manufacturing (p. 57)
Manufacturing usually involves a series of steps, such as designing a product, obtaining and preparing raw materials, processing the materials mechanically or chemically, and assembling the product. All steps may occur at a single location or may occur at different locations. 8B/M2*
In the 1800s, new machinery and steam engines to drive them made it possible to manufacture goods in factories, using fuels as a source of energy. In the factory system, workers, materials, and energy could be brought together efficiently. 10J/M1*
3-5
3C/E1b
Although many things are still made by hand in some parts of the world, almost everything in the most technologically developed countries is now produced using machines that are automated. By using machinery, the time required to make a product and its cost can be greatly reduced. 8B/E4*
In the 1700s, most manufacturing was still done in homes or small shops, using small, handmade machines that were powered by muscle, wind, or moving water. 10J/E1** (BSL)
Manufacturing (p. 57)
8C/E1
Global Interdependence (p. 51)
K-2
technological understanding
history of development
nature of technology

Blanket, Possibly Songhai people, Niger, ca. 1960

*Some important themes pervade science, mathematics, and technology and appear over and over again, whether we are looking at an ancient civilization, the human body, or a comet. They are ideas that transcend disciplinary boundaries and prove fruitful in explanation, in theory, in observation, and in design.*

Science for All Americans

# CHAPTER 11

# COMMON THEMES



Some powerful ideas often used by mathematicians, scientists, and engineers are not the intellectual property of any one field or discipline. Indeed, notions of system, scale, change and constancy, and models have important applications in business and finance, education, law, government and politics, and other domains, as well as in mathematics, science, and technology. These common themes are really ways of thinking rather than theories or discoveries. The ideas mapped in this chapter could be related to almost any field or discipline. Ideas about the physical world, the living environment, and the mathematical world are included here on maps dealing with systems, models, constancy and change, and scale to provide important specific instances of the generalizations.

The list on the left of this page shows the maps for this chapter and indicates the volume of *Atlas of Science Literacy* in which each appears. Those in green typeface are in *Volume 2* and those in gray are in *Volume 1*.

# MODELS (11B)

A model of something is a simplified imitation of it that we hope can help us understand it better. Scientists spend a good deal of time building, testing, comparing and revising models—whether mathematical, physical, or conceptual—and using them to communicate and get ideas about how the real world works. Tables used in determining insurance payments, projections about endangered species, non-destructive testing of bridges, and weather forecasting are all based on models. When a model does not mimic a phenomenon well, the nature of the discrepancy is a clue to how the model can be improved. Models may also mislead, suggesting characteristics that are not really shared with what is being modeled.

The map is organized around two strands—*uses of models* and *limitations of models*. In the elementary grades, the focus is on uses of a variety of models as communication devices and, when the things being modeled are readily observable, on how the models are like and unlike those things. In middle school, the focus is on models of phenomena not accessible through direct observation and on the role of various models, including simulations, in helping us think about phenomena. In high school, the focus is on what can and cannot be learned from models, including computer-based models.

Maps in Chapter 2: THE NATURE OF MATHEMATICS are closely related to the ideas on this map. In particular, the **MATHEMATICAL MODELS** map in *Atlas 1* addresses more extensively the processes of mathematical modeling that are touched on here.

## NOTES

The K-2 benchmark "Many toys are like real things in some ways..." builds on a common understanding of models as three-dimensional miniatures of objects but notes similarities between toys and the real things in terms of what they do in addition to what they look like. New benchmarks at the grades 3-5 and 6-8 levels point out ways in which models may differ from what they represent and the need to consider whether a model's behavior matches key aspects of what is being modeled.

The high-school benchmark 11B/H3 includes two fairly sophisticated ideas: (a) a model can be tested by comparing its predictions to observations, and (b) a close match between predictions based on the model and observations does not mean that another model might not work equally well or better.

Numerous off-map connections highlight the importance of models to various aspects of science and engineering.

## RESEARCH IN *BENCHMARKS*

Students in lower elementary grades have some understanding that models can be used to show how something works, but they believe that perceptual similarity between the model and what it is used to represent is very important when developing or evaluating models (Penner, Giles, Lehrer, & Schauble, 1997). With repeated cycles of modeling and reflection, lower elementary students can focus more on similarities in function and less on perceptual similarities, and upper elementary students can understand the need for symbolic conventions (rather than only physical resemblance) when developing maps, diagrams, and other related display notations (Lehrer & Schauble, 2000).

Prior to instruction, or after traditional instruction, many middle- and high-school students continue to focus on perceptual rather than functional similarities between models and their referents, and think of models predominantly as small copies of real objects (Grosslight, Unger, Jay, & Smith, 1991; Treagust, Chittleborough, & Mamiala, 2002; Schwartz & White, 2005). Consequently, students often interpret models they encounter in school science too literally and unshared attributes between models and their referents are a cause of misunderstanding (Coll, France, & Taylor, 2005; Harrison & Treagust, 1996). Some middle- and high-school students view visual representations such as maps or diagrams as models, but only a few students view representations of ideas or abstract entities as models (Grosslight et al., 1991).

Many middle- and high-school students think that models are useful for visualizing ideas and for communication purposes (Schwarz & White, 2005; Grosslight et al., 1991). Only a few students think that models are useful in developing and testing ideas and that the usefulness of a model can be tested by comparing its implications to actual observations (Grosslight et al., 1991).

Middle-school and high-school students accept the idea that scientists can have more than one model for the same thing (Grosslight et al., 1991). However, having multiple models may mean for them that one could have literally a different view of the same entity, or that one could emphasize different aspects of the entity—omitting or highlighting certain things to provide greater clarity. Students are rarely aware that there could be different models to explain something or to evaluate alternative hypotheses. They find multiple model use in school science confusing and rarely use multiple models to think about phenomena; even if they do, the idea that one model is "right" and "real" persists (Harrison & Treagust, 1996, 2000). Students may know that models can be changed, but changing a model for them means (typical of high-school students) adding new information or (typical of middle-school students) replacing a part that was made wrong (Grosslight et al., 1991).

Developing and evaluating models *combined* with explicit instruction and reflection about the nature of models and modeling for an extended period of time can be effective in helping middle-school students make progress toward the following ideas: Models are not necessarily physical objects but could be conceptual representations that help scientists to predict and explain; there can be multiple models for the same phenomenon; and models are useful in visualization, predicting phenomena, and conducting investigations that are not otherwise possible (Schwarz & White, 2005). The ideas that scientists revise their models in light of new insights or new data and that not all models are of equal value may be harder to develop (Schwarz & White, 2005).

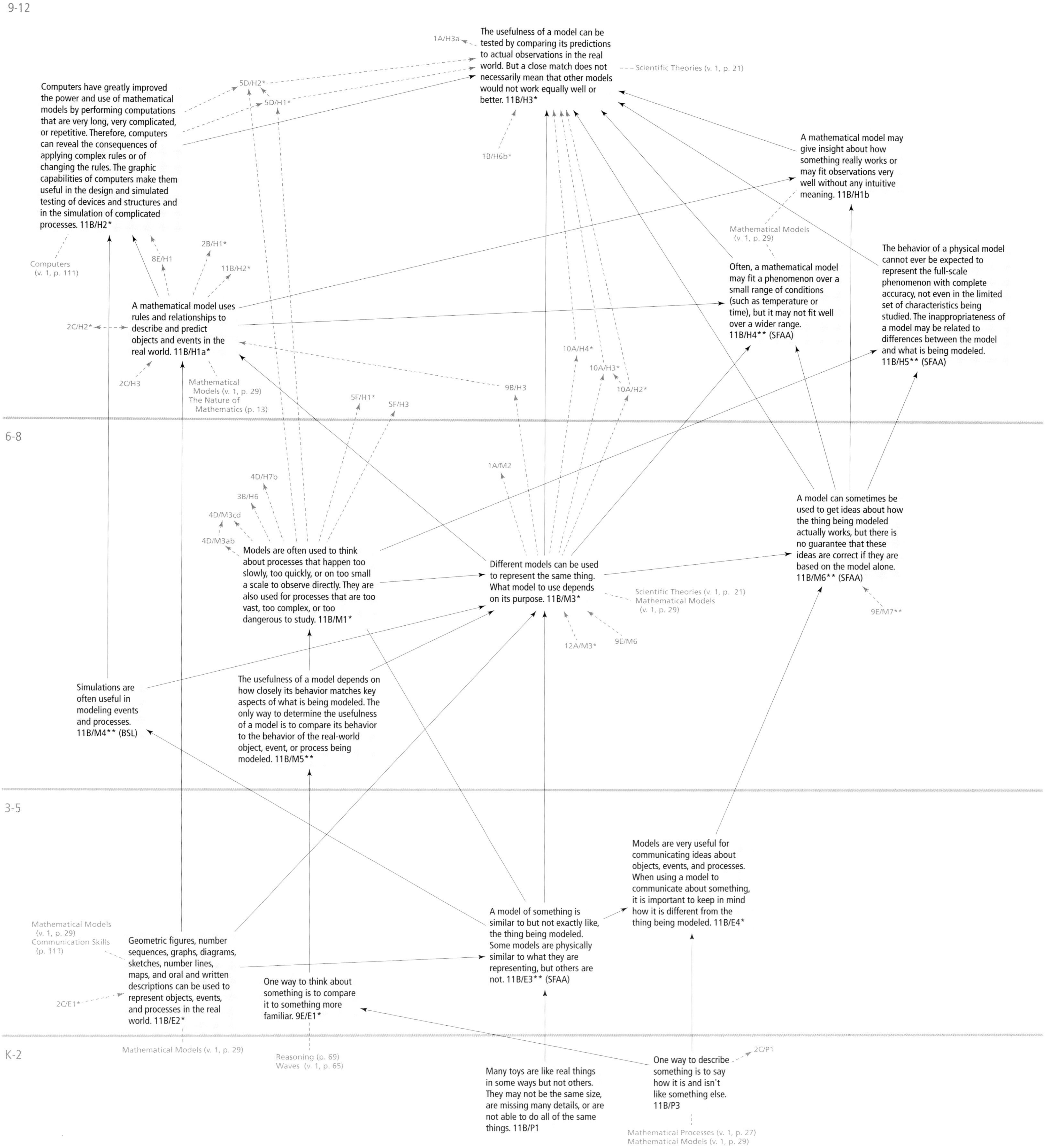


uses of models

limitations of models



conservation

equilibrium

COMMON THEMES

# PATTERNS OF CHANGE (11C)

Much of science and mathematics has to do with understanding how change occurs in nature and in social and technological systems, and much of technology has to do with creating and controlling change. Descriptions of change are important for predicting what will happen; analysis of change is essential for understanding what is going on, as well as for predicting what will happen; and control of change is essential for the design of technological systems.

The map is organized around four strands—*directional change, predicting change, cyclic change,* and *monitoring change*. In the elementary grades, the emphasis is on observable changes and the role of tables and graphs in detecting and displaying patterns of change. In middle school, the focus is on the role of patterns of change in predicting the future. In high school, the focus is on the role of mathematics in detecting change, the role of feedback in cyclic change, and chaos.

The map shows numerous connections to ideas about detectable patterns in natural phenomena, such as ideas found on the **WEATHER AND CLIMATE** map in this volume and on the **BIOLOGICAL EVOLUTION** map in *Atlas 1*. Both of these maps suggest contexts for learning more general ideas about patterns of change.

**NOTES**

The idea of trends as a pattern of change was described in *Science for All Americans* but not developed fully in *Benchmarks for Science Literacy*. This map addresses the idea of trends in the *directional change* strand that includes several new benchmarks: the 3-5 benchmark "Some things, such as...," the 6-8 benchmark "Trends based on what has happened in the past...," and the 9-12 benchmark "Trends that follow a pattern...."

In the *predicting change* strand, two 6-8 benchmarks have been added to explain why it is not always possible to make an accurate prediction and to support 9-12 benchmarks about indeterminacy ("Most systems above the molecular level..." and "The precise future...").

In the *cyclic change* strand, the new 3-5 benchmark "Some things in nature..." is a generalization about repeating patterns of change that summarize examples of cyclic change presented at the K-2 grade range. The new 6-8 benchmark "Cyclic patterns evident in past events..." describes the use and limits of cyclic patterns in making predictions. This idea takes advantage of a similar idea on the use and limits of trends in making predictions, which is presented in the 6-8 benchmark "Trends based on what has happened in the past...."

Because humans are better at noticing patterns when data are presented visually rather than in tables, the *monitoring change* strand focuses primarily on graphic representations of data. The new 9-12 benchmark "It is not always easy..." describes the contribution of statistics in distinguishing meaningful patterns from meaningless ones—sometimes referred to as distinguishing signal from noise—and is based on text in *Science for All Americans.*

**RESEARCH IN *BENCHMARKS***

Fourth-graders' representations of changes over time are "data-driven" in the sense that the particular data in the problem are the most important. This contrasts with "system-driven" representations in which the emphasis is on overall patterns. Unfortunately, students are typically introduced to system-driven representations while they still think it is a wrong or meaningless way to convey information (Tierney & Nemirovsky, 1991).

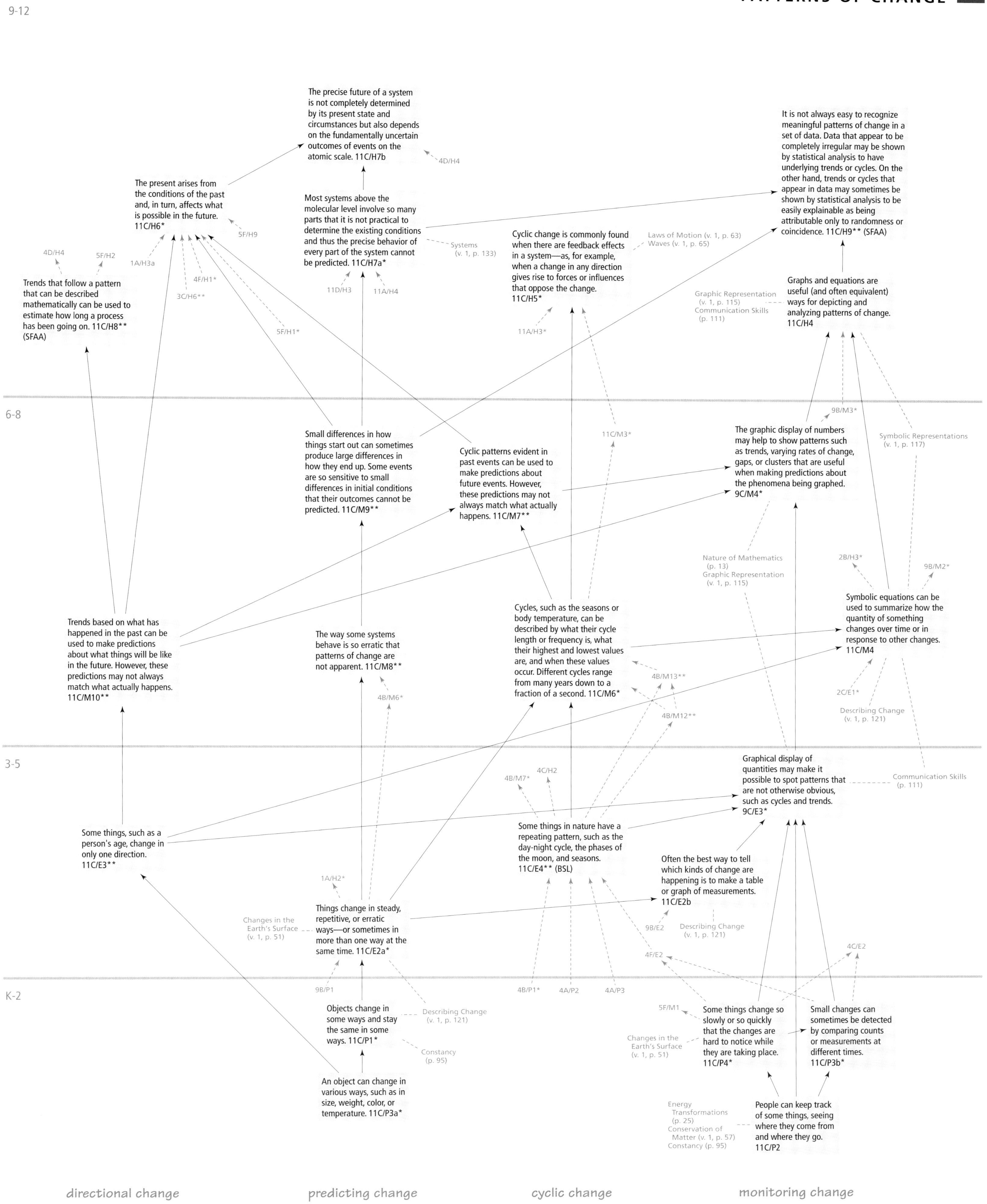
9-12
The precise future of a system is not completely determined by its present state and circumstances but also depends on the fundamentally uncertain outcomes of events on the atomic scale. 11C/H7b
4D/H4
It is not always easy to recognize meaningful patterns of change in a set of data. Data that appear to be completely irregular may be shown by statistical analysis to have underlying trends or cycles. On the other hand, trends or cycles that appear in data may sometimes be shown by statistical analysis to be easily explainable as being attributable only to randomness or coincidence. 11C/H9** (SFAA)
The present arises from the conditions of the past and, in turn, affects what is possible in the future. 11C/H6*
5F/H9
Most systems above the molecular level involve so many parts that it is not practical to determine the existing conditions and thus the precise behavior of every part of the system cannot be predicted. 11C/H7a*
Systems (v. 1, p. 133)
Cyclic change is commonly found when there are feedback effects in a system—as, for example, when a change in any direction gives rise to forces or influences that oppose the change. 11C/H5*
Laws of Motion (v. 1, p. 63)
Waves (v. 1, p. 65)
4D/H4
5F/H2
1A/H3a
4F/H1*
3C/H6**
Trends that follow a pattern that can be described mathematically can be used to estimate how long a process has been going on. 11C/H8** (SFAA)
5F/H1*
11D/H3
11A/H4
11A/H3*
Graphic Representation (v. 1, p. 115)
Communication Skills (p. 111)
Graphs and equations are useful (and often equivalent) ways for depicting and analyzing patterns of change. 11C/H4
6-8
9B/M3*
11C/M3*
Small differences in how things start out can sometimes produce large differences in how they end up. Some events are so sensitive to small differences in initial conditions that their outcomes cannot be predicted. 11C/M9**
Cyclic patterns evident in past events can be used to make predictions about future events. However, these predictions may not always match what actually happens. 11C/M7**
The graphic display of numbers may help to show patterns such as trends, varying rates of change, gaps, or clusters that are useful when making predictions about the phenomena being graphed. 9C/M4*
Symbolic Representations (v. 1, p. 117)
Nature of Mathematics (p. 13)
Graphic Representation (v. 1, p. 115)
2B/H3*
9B/M2*
Trends based on what has happened in the past can be used to make predictions about what things will be like in the future. However, these predictions may not always match what actually happens. 11C/M10**
The way some systems behave is so erratic that patterns of change are not apparent. 11C/M8**
4B/M6*
Cycles, such as the seasons or body temperature, can be described by what their cycle length or frequency is, what their highest and lowest values are, and when these values occur. Different cycles range from many years down to a fraction of a second. 11C/M6*
4B/M13**
4B/M12**
Symbolic equations can be used to summarize how the quantity of something changes over time or in response to other changes. 11C/M4
2C/E1*
Describing Change (v. 1, p. 121)
3-5
4C/H2
4B/M7*
Graphical display of quantities may make it possible to spot patterns that are not otherwise obvious, such as cycles and trends. 9C/E3*
Communication Skills (p. 111)
Some things, such as a person's age, change in only one direction. 11C/E3**
Some things in nature have a repeating pattern, such as the day-night cycle, the phases of the moon, and seasons. 11C/E4** (BSL)
Often the best way to tell which kinds of change are happening is to make a table or graph of measurements. 11C/E2b
1A/H2*
Changes in the Earth's Surface (v. 1, p. 51)
Things change in steady, repetitive, or erratic ways—or sometimes in more than one way at the same time. 11C/E2a*
9B/E2
Describing Change (v. 1, p. 121)
4F/E2
4C/E2
K-2
9B/P1
4B/P1*
4A/P2
4A/P3
Objects change in some ways and stay the same in some ways. 11C/P1*
Describing Change (v. 1, p. 121)
Constancy (p. 95)
5F/M1
Some things change so slowly or so quickly that the changes are hard to notice while they are taking place. 11C/P4*
Changes in the Earth's Surface (v. 1, p. 51)
Small changes can sometimes be detected by comparing counts or measurements at different times. 11C/P3b*
An object can change in various ways, such as in size, weight, color, or temperature. 11C/P3a*
Energy Transformations (p. 25)
Conservation of Matter (v. 1, p. 57)
Constancy (p. 95)
People can keep track of some things, seeing where they come from and where they go. 11C/P2
directional change
predicting change
cyclic change
monitoring change

# SCALE (11D)

The ranges of magnitudes in our universe—size, durations, speeds, and so on—are immense. Nevertheless, we can represent such magnitudes in abstract mathematical terms and seek relationships among them that make sense. Because the properties of an object or parts of a system may be affected differently by changes in size, large changes in size typically change the way that things work in physical, biological, and social systems.

The map is organized around three strands—*effect of size on properties, variety of magnitudes in nature,* and *describing extremes*. The elementary grades focus on familiarity with the range of sizes and properties of familiar objects. In middle school, the range of sizes is expanded to include things as small as atoms and molecules and as big as the universe. In high school, the emphasis is on facility in using powers of ten to describe and compare things and on the recognition that changes in size of an object or system may affect different properties differently.

In addition to being closely related to several maps in Chapter 9: THE MATHEMATICAL WORLD in both *Atlas 1* and *Atlas 2*, this map is also relevant to the **COMPUTATION AND ESTIMATION** map.

## NOTES

At the 9-12 grade range the benchmark "When the linear dimensions..." provides a key example for the *effect of size on properties* strand, focusing on the idea that when something changes size, its volume changes out of proportion to its area. This same benchmark also appears in the **RATIOS AND PROPORTIONALITY** and **DESCRIBING CHANGE** maps in *Atlas 1* where it builds on the idea of two quantities varying but staying in proportion, and in the **SHAPES** map where it builds on ideas about relationships between linear dimensions, area, and volume. This benchmark also contributes to students' understanding of physical models and their limitations, as shown in the **MODELS** map.

A new 6-8 benchmark "Natural phenomena..." addresses the immense range of magnitudes in the universe and the difficulty of appreciating phenomena involving magnitudes far removed from human experience. It builds on students' early experiences of noticing variations in the size, weight, ages, and speeds of things on a human scale. Examples of magnitudes that appear in off-map connections to this benchmark include distances in the universe, the size of atoms and molecules, and the age of the universe.

**RESEARCH IN *BENCHMARKS***

No relevant research available in *Benchmarks*.

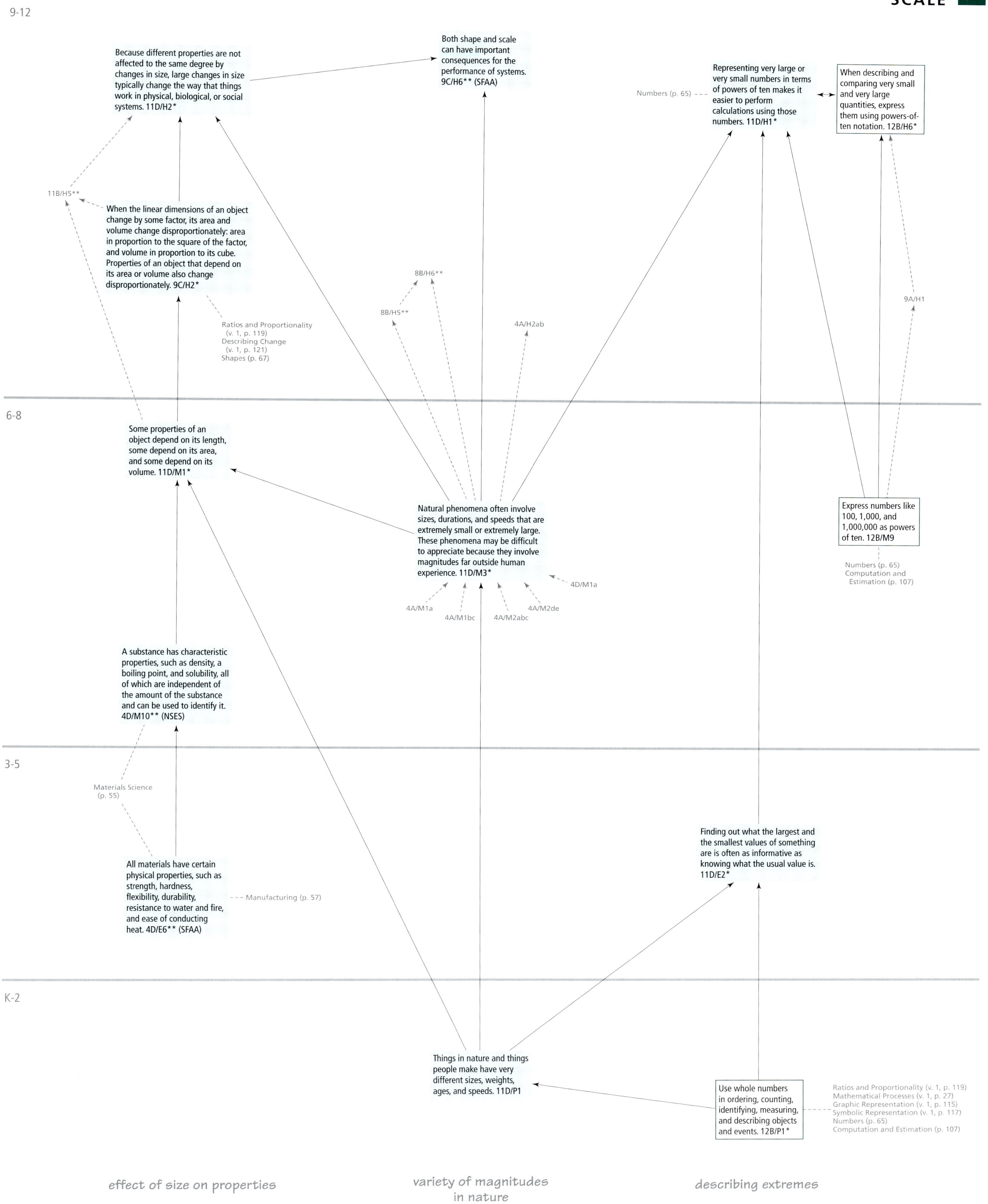
9-12
Because different properties are not affected to the same degree by changes in size, large changes in size typically change the way that things work in physical, biological, or social systems. 11D/H2*
Both shape and scale can have important consequences for the performance of systems. 9C/H6** (SFAA)
Numbers (p. 65)
Representing very large or very small numbers in terms of powers of ten makes it easier to perform calculations using those numbers. 11D/H1*
When describing and comparing very small and very large quantities, express them using powers-of-ten notation. 12B/H6*
11B/H5**
When the linear dimensions of an object change by some factor, its area and volume change disproportionately: area in proportion to the square of the factor, and volume in proportion to its cube. Properties of an object that depend on its area or volume also change disproportionately. 9C/H2*
8B/H6**
8B/H5**
9A/H1
4A/H2ab
Ratios and Proportionality (v. 1, p. 119)
Describing Change (v. 1, p. 121)
Shapes (p. 67)
6-8
Some properties of an object depend on its length, some depend on its area, and some depend on its volume. 11D/M1*
Natural phenomena often involve sizes, durations, and speeds that are extremely small or extremely large. These phenomena may be difficult to appreciate because they involve magnitudes far outside human experience. 11D/M3*
Express numbers like 100, 1,000, and 1,000,000 as powers of ten. 12B/M9
Numbers (p. 65)
Computation and Estimation (p. 107)
4D/M1a
4A/M1a
4A/M1bc
4A/M2abc
4A/M2de
A substance has characteristic properties, such as density, a boiling point, and solubility, all of which are independent of the amount of the substance and can be used to identify it. 4D/M10** (NSES)
3-5
Materials Science (p. 55)
Finding out what the largest and the smallest values of something are is often as informative as knowing what the usual value is. 11D/E2*
All materials have certain physical properties, such as strength, hardness, flexibility, durability, resistance to water and fire, and ease of conducting heat. 4D/E6** (SFAA)
Manufacturing (p. 57)
K-2
Things in nature and things people make have very different sizes, weights, ages, and speeds. 11D/P1
Use whole numbers in ordering, counting, identifying, measuring, and describing objects and events. 12B/P1*
Ratios and Proportionality (v. 1, p. 119)
Mathematical Processes (v. 1, p. 27)
Graphic Representation (v. 1, p. 115)
Symbolic Representation (v. 1, p. 117)
Numbers (p. 65)
Computation and Estimation (p. 107)
effect of size on properties
variety of magnitudes in nature
describing extremes

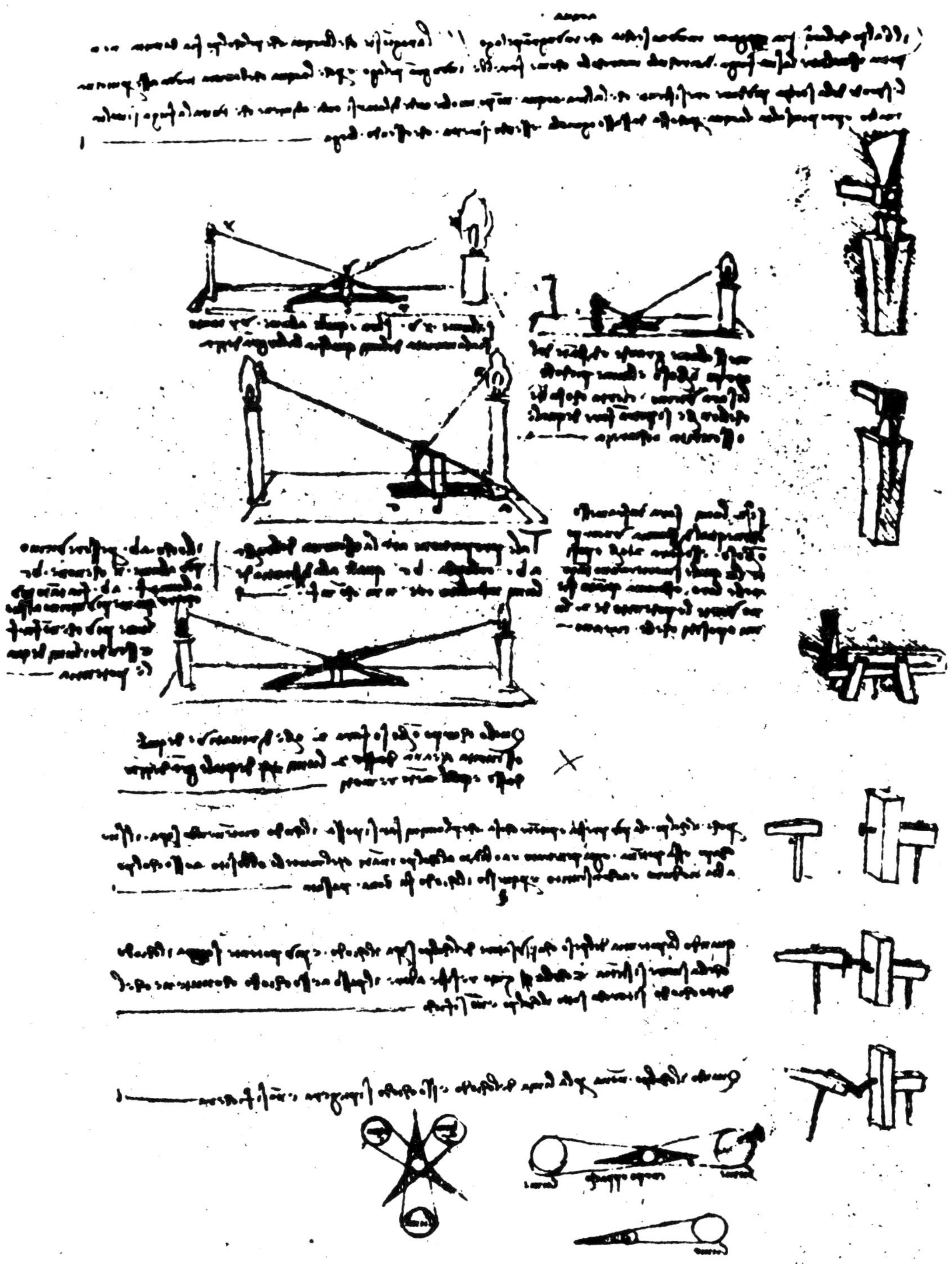

Page from a Leonardo da Vinci notebook, 15th-16th centuries

*Throughout history, people have concerned themselves with the transmission of shared values, attitudes, and skills from one generation to the next. All three were taught long before formal schooling was invented. Even today, it is evident that family, religion, peers, books, news and entertainment media, and general life experiences are the chief influences in shaping people's views of knowledge, learning, and other aspects of life. Science, mathematics, and technology—in the context of schooling—can also play a key role in the process, for they are built upon a distinctive set of values, they reflect and respond to the values of society generally, and they are increasingly influential in shaping shared cultural values. Thus, to the degree that schooling concerns itself with values and attitudes—a matter of great sensitivity in a society that prizes cultural diversity and individuality and is wary of ideology—it must take scientific values and attitudes into account when preparing young people for life beyond school.*

*Similarly, there are certain thinking skills associated with science, mathematics, and technology that young people need to develop during their school years. These are mostly, but not exclusively, mathematical and logical skills that are essential tools for both formal and informal learning and for a lifetime of participation in society as a whole.*

# CHAPTER 12

# HABITS OF MIND



Preparing people to become effective problem solvers, alone and in concert with others, is a major purpose of schooling. Science, mathematics, and technology can contribute significantly to that end because, in their different ways, they are enterprises in the business of searching for solutions to problems ranging from the highly theoretical to the entirely concrete.

Students' ability and inclination to solve problems effectively depend on their having certain knowledge, skills, and attitudes. Quantitative, communication, manual, and critical-response skills are essential for problem solving, but they are also part of what constitutes science literacy more generally. That is why they are brought together as scientific habits of mind.

The list on the left of this page shows the maps for this chapter and indicates the volume of *Atlas of Science Literacy* in which each appears. All of the maps for Chapter 12 are in Volume 2.

# VALUES IN SCIENCE (12A)

To understand the enterprises of science, mathematics, and technology, it is essential to be aware of the values that underlie them and that are shared by the people who work in them: the importance of verifiable data, testable hypotheses, and predictability in science; of rigorous proof and elegance in mathematics; and of optimum design in technology.

Science is also based on everyday values even as it questions our understanding of the world and ourselves. In many respects, science is the systematic application of some highly regarded and widely held human values—integrity, diligence, fairness, curiosity, openness to new ideas, skepticism, and imagination.

The map is organized around two strands—*values particular to science* and *common values as applied in science*. In the elementary grades, the focus is on the importance of observations in making sense of phenomena. In middle school, the focus is on the importance of reproducible evidence and logical reasoning in drawing conclusions from data. In high school, the focus is on understanding why curiosity, honesty, openness, and skepticism are so highly regarded in science and how they are incorporated into the way science is carried out.

The map draws on and contributes to ideas on maps in Chapter 1: THE NATURE OF SCIENCE and to the **DETECTING FLAWS IN ARGUMENTS** map.

## NOTES

The *values particular to science* strand contains three benchmarks at the high-school level that have been developed using ideas found in *Science for All Americans*. Together, they describe the importance of hypotheses, evidence, and theories to the scientific endeavor. In grades 3-5, the new benchmark "Science is a process..." provides young students with an age-appropriate way to understand what science is.

In the *common values as applied in science* strand, two new benchmarks have been developed for the map. Benchmark 12A/E3 provides a reason for students to keep clear and accurate records as is called for in benchmark 12A/E1. At the high-school level, benchmark 12A/H5 provides a justification for students to acquire the traits of curiosity, openness, and skepticism as described in benchmark 12A/H1. Honesty is also called for in benchmark 12A/H1, and the justification for it is already captured in benchmark 1C/H7.

Several benchmarks on the map have been edited to improve their clarity and to better align the expectations on this map with the descriptions of goals in *Science for All Americans.*

**RESEARCH IN *BENCHMARKS***

No relevant research available in *Benchmarks.*

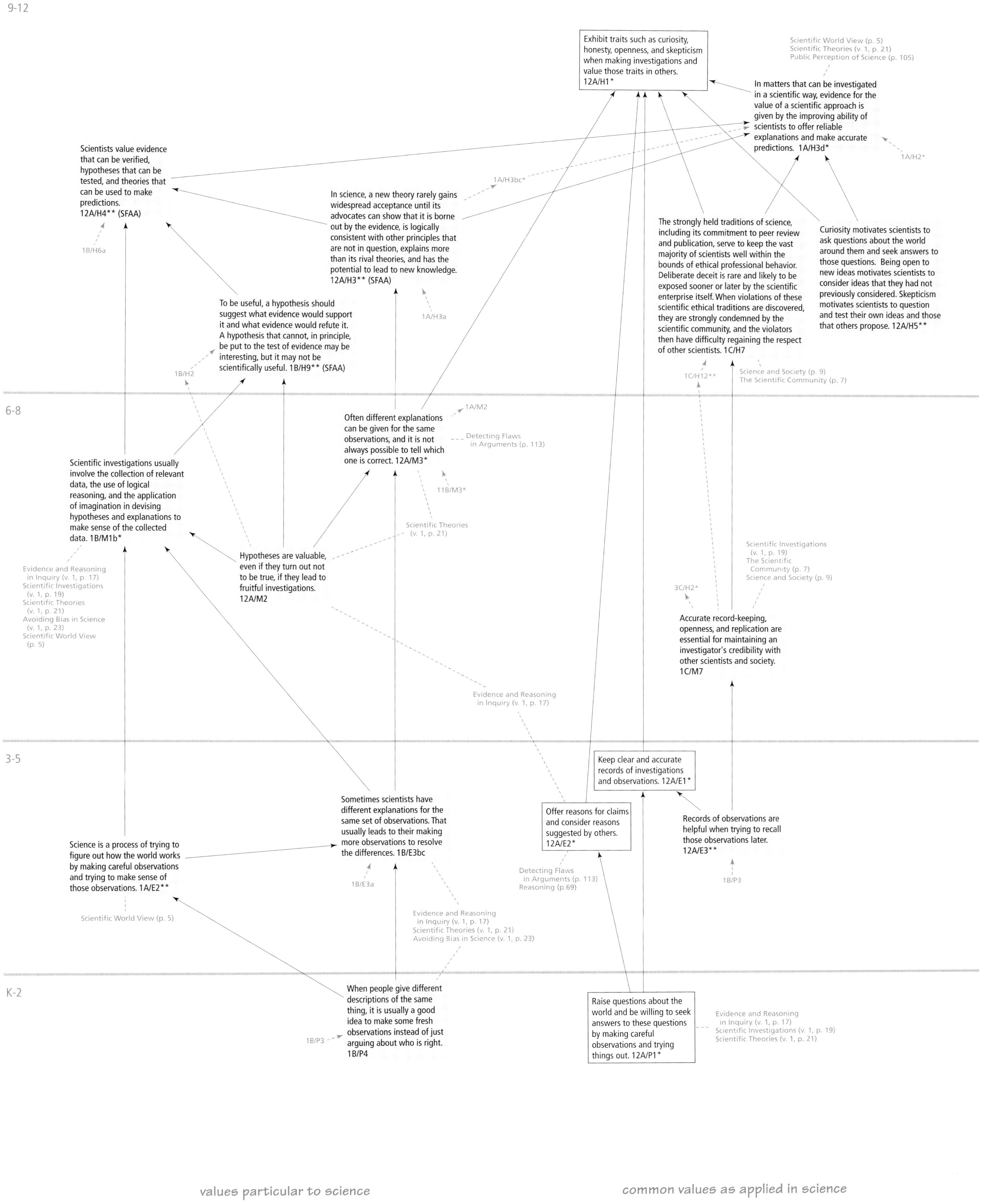
9-12
Exhibit traits such as curiosity, honesty, openness, and skepticism when making investigations and value those traits in others. 12A/H1*
Scientific World View (p. 5)
Scientific Theories (v. 1, p. 21)
Public Perception of Science (p. 105)
In matters that can be investigated in a scientific way, evidence for the value of a scientific approach is given by the improving ability of scientists to offer reliable explanations and make accurate predictions. 1A/H3d*
1A/H2*
Scientists value evidence that can be verified, hypotheses that can be tested, and theories that can be used to make predictions. 12A/H4** (SFAA)
1B/H6a
1A/H3bc*
In science, a new theory rarely gains widespread acceptance until its advocates can show that it is borne out by the evidence, is logically consistent with other principles that are not in question, explains more than its rival theories, and has the potential to lead to new knowledge. 12A/H3** (SFAA)
1A/H3a
The strongly held traditions of science, including its commitment to peer review and publication, serve to keep the vast majority of scientists well within the bounds of ethical professional behavior. Deliberate deceit is rare and likely to be exposed sooner or later by the scientific enterprise itself. When violations of these scientific ethical traditions are discovered, they are strongly condemned by the scientific community, and the violators then have difficulty regaining the respect of other scientists. 1C/H7
Curiosity motivates scientists to ask questions about the world around them and seek answers to those questions. Being open to new ideas motivates scientists to consider ideas that they had not previously considered. Skepticism motivates scientists to question and test their own ideas and those that others propose. 12A/H5**
To be useful, a hypothesis should suggest what evidence would support it and what evidence would refute it. A hypothesis that cannot, in principle, be put to the test of evidence may be interesting, but it may not be scientifically useful. 1B/H9** (SFAA)
1B/H2
1C/H12**
Science and Society (p. 9)
The Scientific Community (p. 7)
6-8
1A/M2
Often different explanations can be given for the same observations, and it is not always possible to tell which one is correct. 12A/M3*
Detecting Flaws in Arguments (p. 113)
11B/M3*
Scientific investigations usually involve the collection of relevant data, the use of logical reasoning, and the application of imagination in devising hypotheses and explanations to make sense of the collected data. 1B/M1b*
Scientific Theories (v. 1, p. 21)
Hypotheses are valuable, even if they turn out not to be true, if they lead to fruitful investigations. 12A/M2
Evidence and Reasoning in Inquiry (v. 1, p. 17)
Scientific Investigations (v. 1, p. 19)
Scientific Theories (v. 1, p. 21)
Avoiding Bias in Science (v. 1, p. 23)
Scientific World View (p. 5)
Scientific Investigations (v. 1, p. 19)
The Scientific Community (p. 7)
Science and Society (p. 9)
3C/H2*
Accurate record-keeping, openness, and replication are essential for maintaining an investigator's credibility with other scientists and society. 1C/M7
Evidence and Reasoning in Inquiry (v. 1, p. 17)
3-5
Keep clear and accurate records of investigations and observations. 12A/E1*
Sometimes scientists have different explanations for the same set of observations. That usually leads to their making more observations to resolve the differences. 1B/E3bc
Offer reasons for claims and consider reasons suggested by others. 12A/E2*
Records of observations are helpful when trying to recall those observations later. 12A/E3**
Science is a process of trying to figure out how the world works by making careful observations and trying to make sense of those observations. 1A/E2**
1B/E3a
Detecting Flaws in Arguments (p. 113)
Reasoning (p.69)
1B/P3
Scientific World View (p. 5)
Evidence and Reasoning in Inquiry (v. 1, p. 17)
Scientific Theories (v. 1, p. 21)
Avoiding Bias in Science (v. 1, p. 23)
K-2
When people give different descriptions of the same thing, it is usually a good idea to make some fresh observations instead of just arguing about who is right. 1B/P4
1B/P3
Raise questions about the world and be willing to seek answers to these questions by making careful observations and trying things out. 12A/P1*
Evidence and Reasoning in Inquiry (v. 1, p. 17)
Scientific Investigations (v. 1, p. 19)
Scientific Theories (v. 1, p. 21)
values particular to science
common values as applied in science

# PUBLIC PERCEPTION OF SCIENCE (12A)

In addition to their positive impact on economic growth, advances in science and technology have profound effects on individuals and societies. Over the years, they have raised both expectations for increased well-being and concerns about possible risks. A goal of science education is an informed public, neither uncritically positive nor categorically antagonistic to science and technology.

The map is organized around three strands—*ethics in research, value and limits of science,* and *value and limits of technology*. In the elementary grades, the focus is on examples of design trade-offs. In middle school, the focus is on impacts of technology on society. High school provides a variety of reasons for holding a balanced perspective of science.

The map draws on and contributes to ideas on maps in Chapter 1: THE NATURE OF SCIENCE and in Chapter 3: THE NATURE OF TECHNOLOGY.

**NOTES**

The main point of this map is described in the high-school benchmark "View science and technology thoughtfully…." All of the other benchmarks can also be found on other maps; their appearance here is to support the formation of students' ideas about the practices and products of science and technology.

The *ethics in research* strand includes benchmarks that are also found on the **SCIENCE AND SOCIETY** map. This includes the high-school benchmark "Because science is a human activity…" that has been derived from *Science for All Americans*.

The *value and limits of science* strand shares several benchmarks with the **SCIENTIFIC WORLD VIEW** map. The four benchmarks at the high-school level are particularly important precursors to viewing science thoughtfully. This includes benchmark 1C/H9, which points out limits on the kinds of questions that science can answer.

Most of the benchmarks in the *value and limit of technology* strand also appear in *Atlas 1* on the **DECISIONS ABOUT TECHNOLOGY** map, the **INTERACTION OF TECHNOLOGY AND SOCIETY** map, or both.

**RESEARCH IN *BENCHMARKS***

No relevant research available in *Benchmarks*.

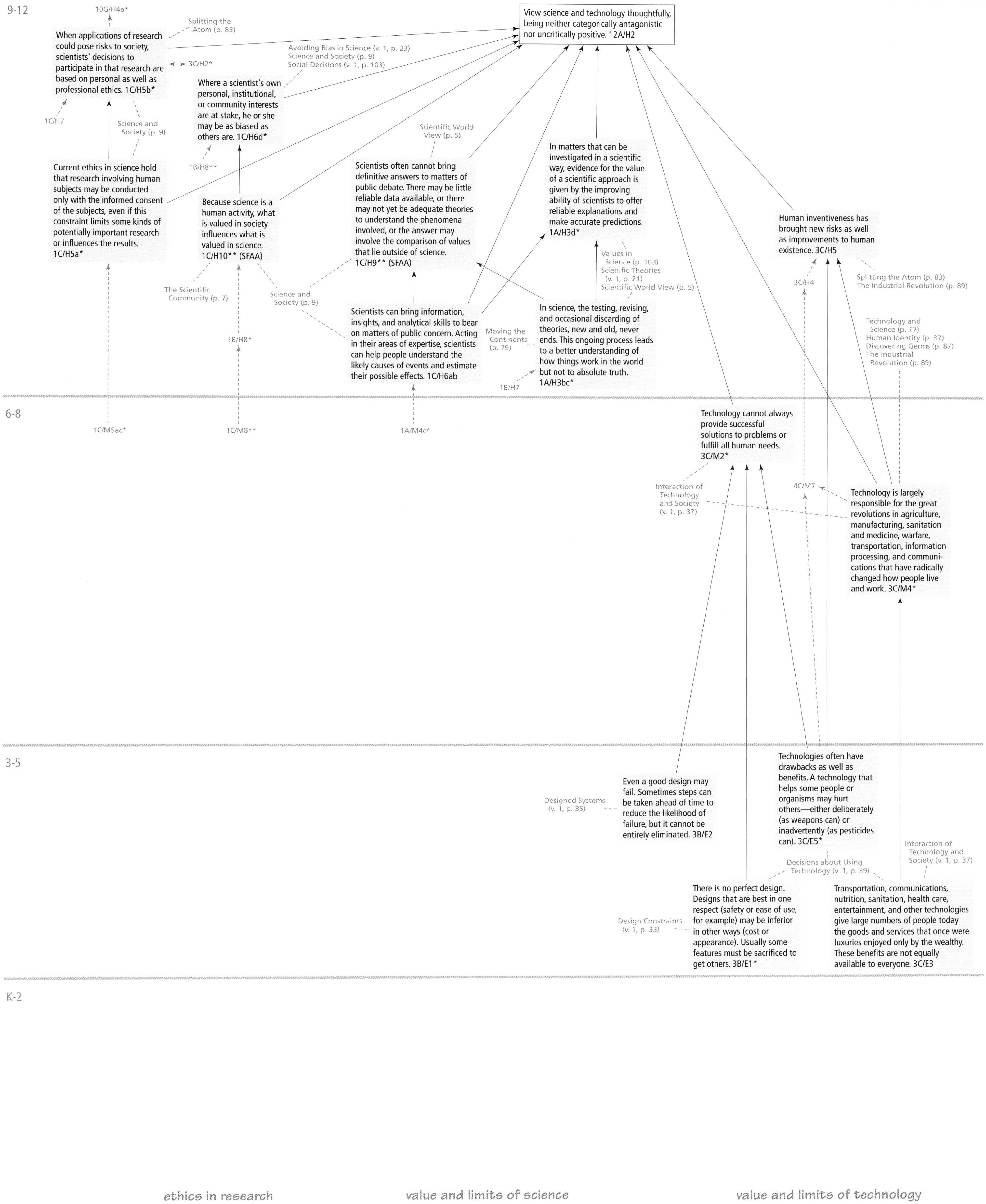
9-12
6-8
3-5
K-2
View science and technology thoughtfully, being neither categorically antagonistic nor uncritically positive. 12A/H2
10G/H4a*
Splitting the Atom (p. 83)
When applications of research could pose risks to society, scientists' decisions to participate in that research are based on personal as well as professional ethics. 1C/H5b*
3C/H2*
1C/H7
Science and Society (p. 9)
Avoiding Bias in Science (v. 1, p. 23)
Science and Society (p. 9)
Social Decisions (v. 1, p. 103)
Where a scientist's own personal, institutional, or community interests are at stake, he or she may be as biased as others are. 1C/H6d*
1B/H8**
Current ethics in science hold that research involving human subjects may be conducted only with the informed consent of the subjects, even if this constraint limits some kinds of potentially important research or influences the results. 1C/H5a*
Because science is a human activity, what is valued in society influences what is valued in science. 1C/H10** (SFAA)
The Scientific Community (p. 7)
Science and Society (p. 9)
1B/H8*
Scientific World View (p. 5)
Scientists often cannot bring definitive answers to matters of public debate. There may be little reliable data available, or there may not yet be adequate theories to understand the phenomena involved, or the answer may involve the comparison of values that lie outside of science. 1C/H9** (SFAA)
In matters that can be investigated in a scientific way, evidence for the value of a scientific approach is given by the improving ability of scientists to offer reliable explanations and make accurate predictions. 1A/H3d*
Values in Science (p. 103)
Scienific Theories (v. 1, p. 21)
Scientific World View (p. 5)
Scientists can bring information, insights, and analytical skills to bear on matters of public concern. Acting in their areas of expertise, scientists can help people understand the likely causes of events and estimate their possible effects. 1C/H6ab
Moving the Continents (p. 79)
In science, the testing, revising, and occasional discarding of theories, new and old, never ends. This ongoing process leads to a better understanding of how things work in the world but not to absolute truth. 1A/H3bc*
1B/H7
Human inventiveness has brought new risks as well as improvements to human existence. 3C/H5
3C/H4
Splitting the Atom (p. 83)
The Industrial Revolution (p. 89)
Technology and Science (p. 17)
Human Identity (p. 37)
Discovering Germs (p. 87)
The Industrial Revolution (p. 89)
1C/M5ac*
1C/M8**
1A/M4c*
Technology cannot always provide successful solutions to problems or fulfill all human needs. 3C/M2*
Interaction of Technology and Society (v. 1, p. 37)
4C/M7
Technology is largely responsible for the great revolutions in agriculture, manufacturing, sanitation and medicine, warfare, transportation, information processing, and communications that have radically changed how people live and work. 3C/M4*
Even a good design may fail. Sometimes steps can be taken ahead of time to reduce the likelihood of failure, but it cannot be entirely eliminated. 3B/E2
Designed Systems (v. 1, p. 35)
Technologies often have drawbacks as well as benefits. A technology that helps some people or organisms may hurt others—either deliberately (as weapons can) or inadvertently (as pesticides can). 3C/E5*
Decisions about Using Technology (v. 1, p. 39)
Interaction of Technology and Society (v. 1, p. 37)
There is no perfect design. Designs that are best in one respect (safety or ease of use, for example) may be inferior in other ways (cost or appearance). Usually some features must be sacrificed to get others. 3B/E1*
Design Constraints (v. 1, p. 33)
Transportation, communications, nutrition, sanitation, health care, entertainment, and other technologies give large numbers of people today the goods and services that once were luxuries enjoyed only by the wealthy. These benefits are not equally available to everyone. 3C/E3
ethics in research
value and limits of science
value and limits of technology

# COMPUTATION AND ESTIMATION (12B)

In everyday life, people often have to make simple calculations in their minds. The advent of the small, inexpensive electronic calculator has made it possible to carry out a range of calculations with precision and to check the reasonableness of the answers. There are also many situations where an approximate answer is as useful as a precise one. Estimating approximate answers can often take the place of making a precise measurement or a careful calculation or be used to check calculations made using electronic calculators or paper and pencil. Skill in estimation is based on a sense of what level of precision is needed, which in turn depends on understanding the purpose of the calculation.

The map is organized around two strands—*estimation* and *computation*. In elementary grades, the focus is on using numbers to represent quantities and basic operations on numbers. In middle school, the kinds of numbers involved in problems expands to include fractions, decimals, and percents and the kinds of computations expand to include calculation of central tendency and probability. Computer spreadsheets come into play as a way to perform arithmetic operations on numbers. In high school, problem solving includes use of ratio and proportion, simple algebraic formulas, and computer-based analysis. Throughout, students are expected to make sound choices about which tools to use—their heads, paper and pencil, calculators, or computers—and to judge the reasonableness of measurements and calculations.

This map draws on mathematical ideas that also appear on the **MATHEMATICAL PROCESSES** map in *Atlas 1* and the **NUMBERS** map in this volume.

## NOTES

The grades 3-5 benchmark "Make calculations when necessary to solve real-world problems..." is central to the overall topic of the map, as indicated by the number of arrows connecting it to other benchmarks. The link between this benchmark and the benchmark "State the purpose of each step in a calculation" indicates that it is not sufficient for students to be able to solve problems simply by memorizing steps in a solution. They are also expected to understand the purpose of each step in all of the calculations they make. Three new benchmarks derived from *Science for All Americans* identify specific skills needed for solving a wide range of problems.

"EVERY ONCE IN A WHILE I JUST LIKE TO UNWIND WITH A LITTLE ADDITION AND SUBTRACTION."

### RESEARCH IN *BENCHMARKS*

Research suggests using word problems as a basis for teaching addition and subtraction concepts, rather than teaching computational skills first and then applying them to solve problems (Carpenter & Moser, 1983). Research has identified a developmental progression of concepts and skills that students use for addition and subtraction (Fuson, 1988; Fuson, 1992). There is evidence that instruction based on this progression can help (Romberg & Carpenter, 1986; Fuson & Willis, 1989).

Students make a variety of errors in multi-digit addition and subtraction calculations (Brown & Van Lehn, 1982). Student errors suggest students interpret and treat multi-digit numbers as single-digit numbers placed adjacent to each other, rather than using place-value meanings for the digits in different positions (Fuson, 1992). Research also suggests students interpret multiplication of whole numbers mainly as repeated addition. This interpretation is inadequate for many multiplication problems and can lead to restrictive intuitive notions such as "multiplication always makes larger"(Greer, 1992).

Elementary- and middle-school students make several errors when they operate on decimals and fractions (Benander & Clement, 1985; Kouba et al., 1988; Peck & Jencks, 1981; Wearne & Hiebert, 1988). For example, many middle-school students cannot add 4 + 0.3 correctly or $7\frac{1}{6} + 3\frac{1}{2}$ (Kouba et al., 1988; Wearne & Hiebert, 1988). Also, lower middle-school students may have difficulties understanding the relationship between factions and decimals (Markovits & Sowder, 1991). These errors are due in part to the fact that students lack essential concepts about decimals and fractions and have memorized procedures that they apply incorrectly. Interventions to improve concept knowledge can lead to increased ability by $5^{th}$-graders to add and subtract decimals correctly (Wearne & Hiebert, 1988).

Students of all ages misunderstand multiplication and division (Bell et al., 1984; Graeber & Tirosh, 1988; Greer, 1992). Commonly held misconceptions include "multiplication always makes larger" and "division always makes smaller." Students may correctly select multiplication as the operation needed to calculate the cost of gasoline when the amount and unit cost are integers, then select division for the same problem when the amount and unit cost are decimal numbers (Bell et al., 1981). Suggestions have been made to improve student concepts of multiplication (Greer, 1992), but further research is needed to evaluate these suggestions.

The use of calculators in K-12 mathematics does not hinder the development of basic computation skills and frequently improves concept development and paper-and-pencil skills, both in basic operations and in problem solving (Hembree & Dessart, 1986; Kaput, 1992). The use of calculators in testing produces higher scores than paper-and-pencil efforts in problem solving as well as in basic operations (Hembree & Dessart, 1986).

Good estimators use a variety of estimating tactics and switch easily between them. They understand place value and the meaning of operations, and are skilled in mental computation. Poor estimators rely on algorithms that are more likely to yield the exact answer. They lack an understanding of the notion and value of estimation and often describe it as "guessing" (Sowder, 1992b). Before 6th grade, students develop few estimation skills from their natural experiences (Case & Sowder, 1990; Sowder, 1992b). Some researchers caution that teaching estimation to young children may have as its single effect that they master specific procedures in a superficial manner (Sowder, 1992b).

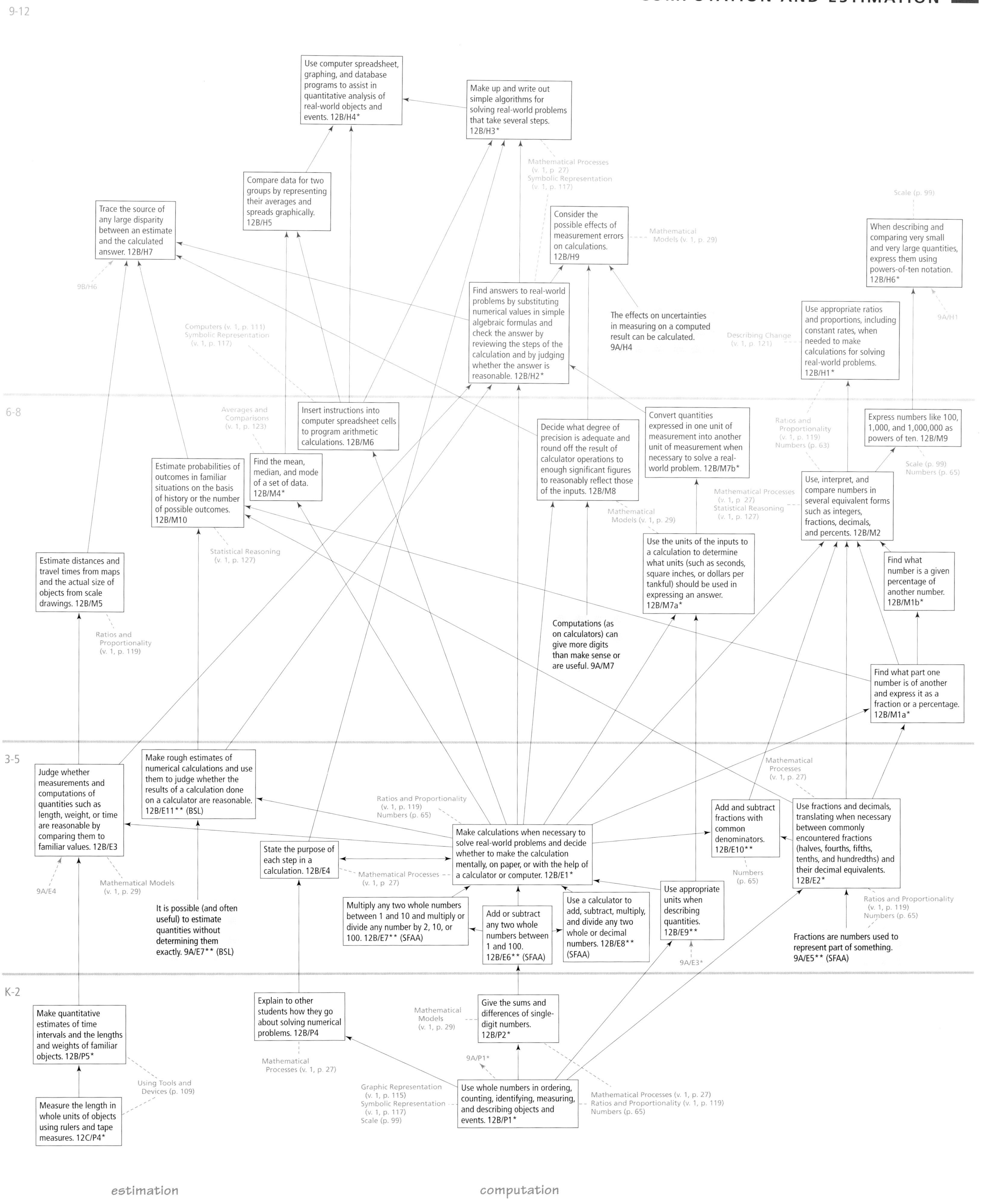
9-12
Use computer spreadsheet, graphing, and database programs to assist in quantitative analysis of real-world objects and events. 12B/H4*
Make up and write out simple algorithms for solving real-world problems that take several steps. 12B/H3*
Mathematical Processes (v. 1, p. 27)
Symbolic Representation (v. 1, p. 117)
Compare data for two groups by representing their averages and spreads graphically. 12B/H5
Trace the source of any large disparity between an estimate and the calculated answer. 12B/H7
Consider the possible effects of measurement errors on calculations. 12B/H9
Mathematical Models (v. 1, p. 29)
Scale (p. 99)
When describing and comparing very small and very large quantities, express them using powers-of-ten notation. 12B/H6*
9B/H6
Find answers to real-world problems by substituting numerical values in simple algebraic formulas and check the answer by reviewing the steps of the calculation and by judging whether the answer is reasonable. 12B/H2*
Computers (v. 1, p. 111)
Symbolic Representation (v. 1, p. 117)
The effects on uncertainties in measuring on a computed result can be calculated. 9A/H4
Describing Change (v. 1, p. 121)
Use appropriate ratios and proportions, including constant rates, when needed to make calculations for solving real-world problems. 12B/H1*
9A/H1
6-8
Averages and Comparisons (v. 1, p. 123)
Insert instructions into computer spreadsheet cells to program arithmetic calculations. 12B/M6
Decide what degree of precision is adequate and round off the result of calculator operations to enough significant figures to reasonably reflect those of the inputs. 12B/M8
Convert quantities expressed in one unit of measurement into another unit of measurement when necessary to solve a real-world problem. 12B/M7b*
Ratios and Proportionality (v. 1, p. 119)
Numbers (p. 63)
Express numbers like 100, 1,000, and 1,000,000 as powers of ten. 12B/M9
Estimate probabilities of outcomes in familiar situations on the basis of history or the number of possible outcomes. 12B/M10
Find the mean, median, and mode of a set of data. 12B/M4*
Scale (p. 99)
Numbers (p. 65)
Mathematical Processes (v. 1, p. 27)
Statistical Reasoning (v. 1, p. 127)
Use, interpret, and compare numbers in several equivalent forms such as integers, fractions, decimals, and percents. 12B/M2
Mathematical Models (v. 1, p. 29)
Statistical Reasoning (v. 1, p. 127)
Estimate distances and travel times from maps and the actual size of objects from scale drawings. 12B/M5
Use the units of the inputs to a calculation to determine what units (such as seconds, square inches, or dollars per tankful) should be used in expressing an answer. 12B/M7a*
Find what number is a given percentage of another number. 12B/M1b*
Ratios and Proportionality (v. 1, p. 119)
Computations (as on calculators) can give more digits than make sense or are useful. 9A/M7
Find what part one number is of another and express it as a fraction or a percentage. 12B/M1a*
3-5
Make rough estimates of numerical calculations and use them to judge whether the results of a calculation done on a calculator are reasonable. 12B/E11** (BSL)
Judge whether measurements and computations of quantities such as length, weight, or time are reasonable by comparing them to familiar values. 12B/E3
Ratios and Proportionality (v. 1, p. 119)
Numbers (p. 65)
Mathematical Processes (v. 1, p. 27)
Add and subtract fractions with common denominators. 12B/E10**
Use fractions and decimals, translating when necessary between commonly encountered fractions (halves, fourths, fifths, tenths, and hundredths) and their decimal equivalents. 12B/E2*
State the purpose of each step in a calculation. 12B/E4
Make calculations when necessary to solve real-world problems and decide whether to make the calculation mentally, on paper, or with the help of a calculator or computer. 12B/E1*
Mathematical Processes (v. 1, p. 27)
Numbers (p. 65)
9A/E4
Mathematical Models (v. 1, p. 29)
Multiply any two whole numbers between 1 and 10 and multiply or divide any number by 2, 10, or 100. 12B/E7** (SFAA)
Add or subtract any two whole numbers between 1 and 100. 12B/E6** (SFAA)
Use a calculator to add, subtract, multiply, and divide any two whole or decimal numbers. 12B/E8** (SFAA)
Use appropriate units when describing quantities. 12B/E9**
Ratios and Proportionality (v. 1, p. 119)
Numbers (p. 65)
It is possible (and often useful) to estimate quantities without determining them exactly. 9A/E7** (BSL)
9A/E3*
Fractions are numbers used to represent part of something. 9A/E5** (SFAA)
K-2
Make quantitative estimates of time intervals and the lengths and weights of familiar objects. 12B/P5*
Explain to other students how they go about solving numerical problems. 12B/P4
Mathematical Models (v. 1, p. 29)
Give the sums and differences of single-digit numbers. 12B/P2*
Mathematical Processes (v. 1, p. 27)
9A/P1*
Using Tools and Devices (p. 109)
Graphic Representation (v. 1, p. 115)
Symbolic Representation (v. 1, p. 117)
Scale (p. 99)
Use whole numbers in ordering, counting, identifying, measuring, and describing objects and events. 12B/P1*
Mathematical Processes (v. 1, p. 27)
Ratios and Proportionality (v. 1, p. 119)
Numbers (p. 65)
Measure the length in whole units of objects using rulers and tape measures. 12C/P4*
estimation
computation

# USING TOOLS AND DEVICES (12C)

Everyone should acquire the ability to handle common materials and tools for dealing with household and other everyday technologies, for making careful observations, and for managing information.

The map is organized around three strands—*making accurate measurements, working with tools and troubleshooting devices,* and *recording and organizing information*. In the elementary grades, the focus is on measuring familiar objects, making something functional, and recording information. In middle school, the emphasis is on selecting appropriate tools and devices for measurement and construction and using computers to store and retrieve information. In high school, the emphasis is on learning how to use and troubleshoot mechanical and electrical devices and on developing simple databases to store and retrieve information.

The map draws on ideas about storing and processing information found on the **COMPUTERS** map in *Atlas 1* and on ideas about shapes and their relationships found on the **SHAPES** map in this volume.

## NOTES

In the *making accurate measurements* strand, the benchmark "Weigh objects using a scale" has been added to give students first-hand experience in measuring the weights of objects before they are asked to make estimates of weights (as in benchmark 12B/P5).

In the *working with tools and troubleshooting devices* strand, the grades 6-8 benchmark "Make safe electrical connections...," which had been placed at the 3-5 level in *Benchmarks for Science Literacy*, has been moved to the middle-school level where it joins other benchmarks about electric circuits. Two new middle-school benchmarks about the selection and care of tools also have been added to this strand.

The prevalence of technologies to manage information has led to the addition or revision of several benchmarks in the *recording and organizing information* strand. For example, the use of databases has become so common that the term has been added to a middle-school benchmark (12C/M2), and a new benchmark "Develop simple computer databases..." has been added to the high-school level. In addition, audio and video recorders have become so simple to use that the skills originally expected of middle-school students have been moved to the elementary level (12C/E6).

## RESEARCH IN *BENCHMARKS*

Upper elementary- and middle-school students who can use measuring instruments and procedures when asked to do so often do not use this ability while performing an investigation. Typically a student asked to undertake an investigation and given a set of equipment that includes measuring instruments will make a qualitative comparison even though she might be competent to use the instruments in a different context (Black, 1990). It appears students often know how to take measurements but not what measurements to make or when.

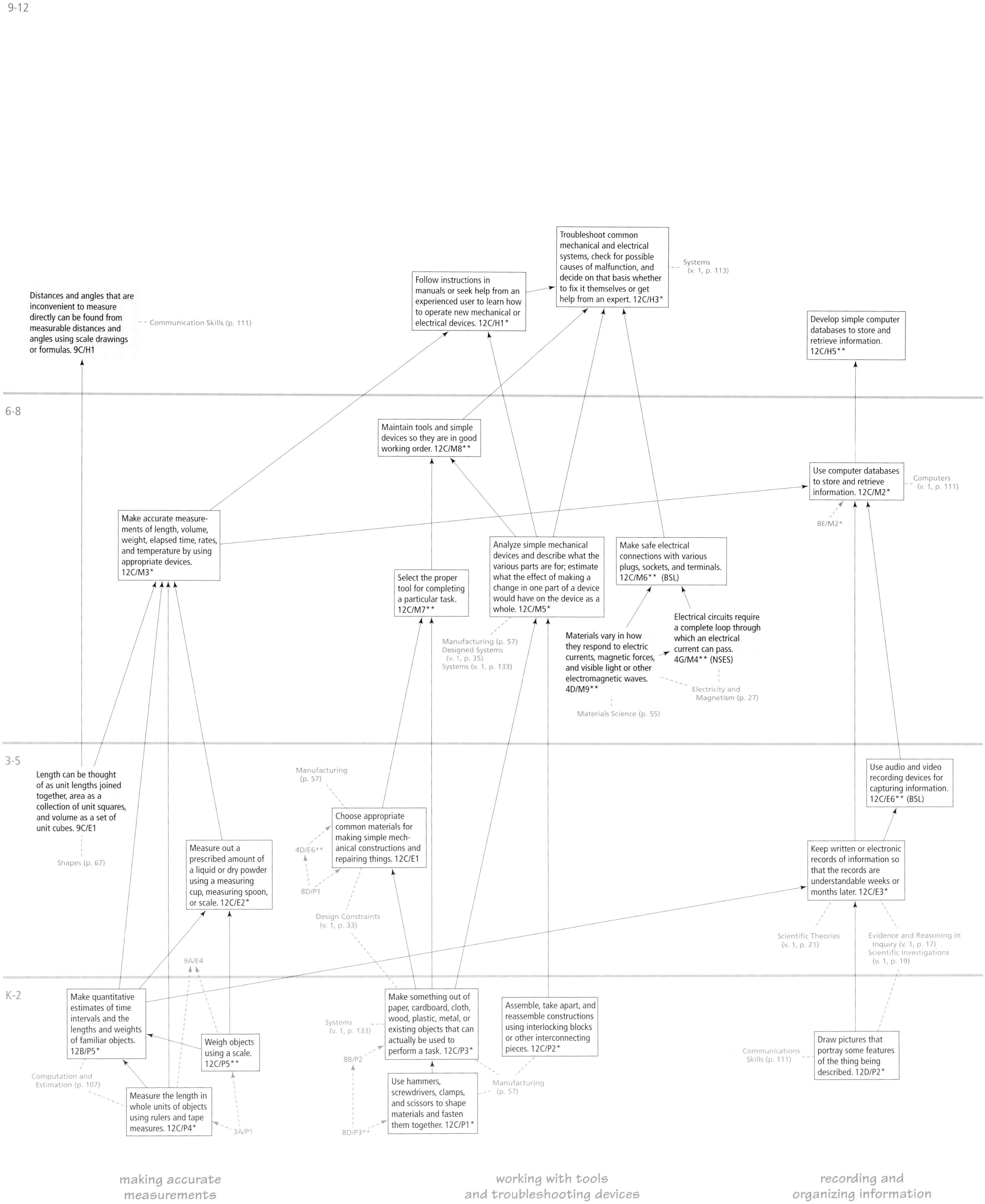
9-12
6-8
3-5
K-2
Distances and angles that are inconvenient to measure directly can be found from measurable distances and angles using scale drawings or formulas. 9C/H1
Communication Skills (p. 111)
Follow instructions in manuals or seek help from an experienced user to learn how to operate new mechanical or electrical devices. 12C/H1*
Troubleshoot common mechanical and electrical systems, check for possible causes of malfunction, and decide on that basis whether to fix it themselves or get help from an expert. 12C/H3*
Systems (v. 1, p. 113)
Develop simple computer databases to store and retrieve information. 12C/H5**
Maintain tools and simple devices so they are in good working order. 12C/M8**
Use computer databases to store and retrieve information. 12C/M2*
Computers (v. 1, p. 111)
8E/M2*
Make accurate measurements of length, volume, weight, elapsed time, rates, and temperature by using appropriate devices. 12C/M3*
Select the proper tool for completing a particular task. 12C/M7**
Analyze simple mechanical devices and describe what the various parts are for; estimate what the effect of making a change in one part of a device would have on the device as a whole. 12C/M5*
Make safe electrical connections with various plugs, sockets, and terminals. 12C/M6** (BSL)
Manufacturing (p. 57)
Designed Systems (v. 1, p. 35)
Systems (v. 1, p. 133)
Materials vary in how they respond to electric currents, magnetic forces, and visible light or other electromagnetic waves. 4D/M9**
Electrical circuits require a complete loop through which an electrical current can pass. 4G/M4** (NSES)
Electricity and Magnetism (p. 27)
Materials Science (p. 55)
Length can be thought of as unit lengths joined together, area as a collection of unit squares, and volume as a set of unit cubes. 9C/E1
Shapes (p. 67)
Measure out a prescribed amount of a liquid or dry powder using a measuring cup, measuring spoon, or scale. 12C/E2*
Manufacturing (p. 57)
Choose appropriate common materials for making simple mechanical constructions and repairing things. 12C/E1
4D/E6**
8D/P1
Design Constraints (v. 1, p. 33)
Use audio and video recording devices for capturing information. 12C/E6** (BSL)
Keep written or electronic records of information so that the records are understandable weeks or months later. 12C/E3*
Scientific Theories (v. 1, p. 21)
Evidence and Reasoning in Inquiry (v. 1, p. 17)
Scientific Investigations (v. 1, p. 19)
9A/E4
Make quantitative estimates of time intervals and the lengths and weights of familiar objects. 12B/P5*
Weigh objects using a scale. 12C/P5**
Computation and Estimation (p. 107)
Measure the length in whole units of objects using rulers and tape measures. 12C/P4*
3A/P1
Systems (v. 1, p. 133)
Make something out of paper, cardboard, cloth, wood, plastic, metal, or existing objects that can actually be used to perform a task. 12C/P3*
8B/P2
Use hammers, screwdrivers, clamps, and scissors to shape materials and fasten them together. 12C/P1*
8D/P3**
Assemble, take apart, and reassemble constructions using interlocking blocks or other interconnecting pieces. 12C/P2*
Manufacturing (p. 57)
Communications Skills (p. 111)
Draw pictures that portray some features of the thing being described. 12D/P2*
making accurate measurements
working with tools and troubleshooting devices
recording and organizing information

# COMMUNICATION SKILLS (12D)

Discourse in science, mathematics, and technology calls for the ability to communicate ideas and share information with fidelity and clarity and to read and listen with understanding. Some of the skills involved are specific to science, mathematics, and technology, and others are general—although even those are not independent of content.

The map is organized around three strands—*mathematical communication, visual communication,* and *oral and written communication*. In the elementary grades, the focus is on describing and interpreting descriptions of objects and events, including descriptions using simple tables and graphs. In middle school, the focus is on interpreting graphic and symbolic representations of data and on giving scientific explanations of phenomena. In high school, the focus is on using graphs and equations to represent relationships among objects and events and to make logical arguments about claims.

The map draws on ideas about shapes and their relationships on the **SHAPES** map and ideas about monitoring change on the **PATTERNS OF CHANGE** map.

### NOTES

Several benchmarks have been added to this map to provide a more coherent progression of skills over the course of a K-12 education. In the *mathematical communication* strand, these include a new grades 3-5 benchmark "Read simple tables and graphs..." along with new benchmarks on interpreting simple symbolic equations at the middle-school level and on developing symbolic equations at the high-school level.

A benchmark about reading maps and globes in the *visual communication* strand has been moved from the 6-8 to 3-5 grade range to better match expectations in the national standards for social studies and geography. In grades K-2, a new statement "Interpret pictures, drawings, and videos..." has been added as a precursor to a 3-5 benchmark that expects students to make their own accurate representations. A new 6-8 benchmark "Prepare a visual presentation..." calls for students to make use of visualizations in their explanations.

In the *oral and written communication* strand, new benchmarks ask students to interpret oral descriptions in grades K-2 and written descriptions in grades 3-5. The elementary-level benchmark "Locate information in print and electronic resources" describes skills that have become more prevalent among younger students than they were at the time the skills were originally placed at the middle-school level in *Benchmarks for Science Literacy*. Finally, to prepare for the high-school benchmark "Participate in group discussions on scientific topics...," three new prerequisite benchmarks have been added at the grades 6-8 level, including "Present a brief scientific explanation orally or in writing..." and "Seek to gain a better understanding...."

### RESEARCH IN *BENCHMARKS*

No relevant research available in *Benchmarks*.

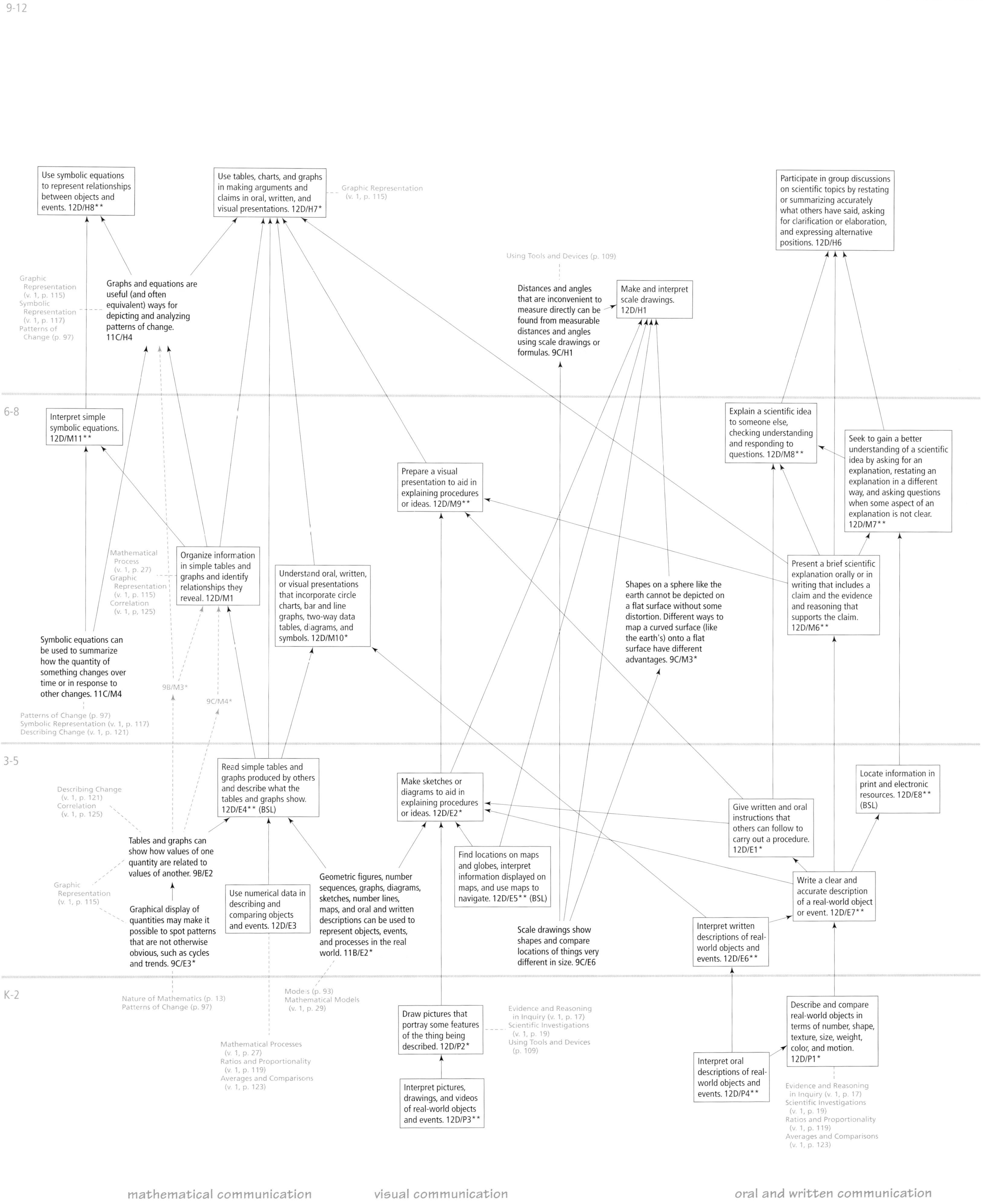
9-12
Use symbolic equations to represent relationships between objects and events. 12D/H8**
Use tables, charts, and graphs in making arguments and claims in oral, written, and visual presentations. 12D/H7*
Graphic Representation (v. 1, p. 115)
Participate in group discussions on scientific topics by restating or summarizing accurately what others have said, asking for clarification or elaboration, and expressing alternative positions. 12D/H6
Using Tools and Devices (p. 109)
Graphic Representation (v. 1, p. 115) Symbolic Representation (v. 1, p. 117) Patterns of Change (p. 97)
Graphs and equations are useful (and often equivalent) ways for depicting and analyzing patterns of change. 11C/H4
Distances and angles that are inconvenient to measure directly can be found from measurable distances and angles using scale drawings or formulas. 9C/H1
Make and interpret scale drawings. 12D/H1
6-8
Interpret simple symbolic equations. 12D/M11**
Explain a scientific idea to someone else, checking understanding and responding to questions. 12D/M8**
Seek to gain a better understanding of a scientific idea by asking for an explanation, restating an explanation in a different way, and asking questions when some aspect of an explanation is not clear. 12D/M7**
Prepare a visual presentation to aid in explaining procedures or ideas. 12D/M9**
Mathematical Process (v. 1, p. 27) Graphic Representation (v. 1, p. 115) Correlation (v. 1, p. 125)
Organize information in simple tables and graphs and identify relationships they reveal. 12D/M1
Understand oral, written, or visual presentations that incorporate circle charts, bar and line graphs, two-way data tables, diagrams, and symbols. 12D/M10*
Present a brief scientific explanation orally or in writing that includes a claim and the evidence and reasoning that supports the claim. 12D/M6**
Shapes on a sphere like the earth cannot be depicted on a flat surface without some distortion. Different ways to map a curved surface (like the earth's) onto a flat surface have different advantages. 9C/M3*
Symbolic equations can be used to summarize how the quantity of something changes over time or in response to other changes. 11C/M4
9B/M3*
9C/M4*
Patterns of Change (p. 97) Symbolic Representation (v. 1, p. 117) Describing Change (v. 1, p. 121)
3-5
Describing Change (v. 1, p. 121) Correlation (v. 1, p. 125)
Read simple tables and graphs produced by others and describe what the tables and graphs show. 12D/E4** (BSL)
Make sketches or diagrams to aid in explaining procedures or ideas. 12D/E2*
Locate information in print and electronic resources. 12D/E8** (BSL)
Give written and oral instructions that others can follow to carry out a procedure. 12D/E1*
Tables and graphs can show how values of one quantity are related to values of another. 9B/E2
Graphic Representation (v. 1, p. 115)
Graphical display of quantities may make it possible to spot patterns that are not otherwise obvious, such as cycles and trends. 9C/E3*
Use numerical data in describing and comparing objects and events. 12D/E3
Geometric figures, number sequences, graphs, diagrams, sketches, number lines, maps, and oral and written descriptions can be used to represent objects, events, and processes in the real world. 11B/E2*
Find locations on maps and globes, interpret information displayed on maps, and use maps to navigate. 12D/E5** (BSL)
Write a clear and accurate description of a real-world object or event. 12D/E7**
Scale drawings show shapes and compare locations of things very different in size. 9C/E6
Interpret written descriptions of real-world objects and events. 12D/E6**
K-2
Nature of Mathematics (p. 13) Patterns of Change (p. 97)
Models (p. 93) Mathematical Models (v. 1, p. 29)
Draw pictures that portray some features of the thing being described. 12D/P2*
Evidence and Reasoning in Inquiry (v. 1, p. 17) Scientific Investigations (v. 1, p. 19) Using Tools and Devices (p. 109)
Describe and compare real-world objects in terms of number, shape, texture, size, weight, color, and motion. 12D/P1*
Mathematical Processes (v. 1, p. 27) Ratios and Proportionality (v. 1, p. 119) Averages and Comparisons (v. 1, p. 123)
Interpret pictures, drawings, and videos of real-world objects and events. 12D/P3**
Interpret oral descriptions of real-world objects and events. 12D/P4**
Evidence and Reasoning in Inquiry (v. 1, p. 17) Scientific Investigations (v. 1, p. 19) Ratios and Proportionality (v. 1, p. 119) Averages and Comparisons (v. 1, p. 123)
mathematical communication
visual communication
oral and written communication

# DETECTING FLAWS IN ARGUMENTS (12E)

In various forms, the mass media, authority figures, and peers inundate us with assertions and arguments, some of them in the realm of science, mathematics, and technology. Education should prepare people to read or listen to such assertions critically, deciding what evidence to pay attention to and what to dismiss, and distinguishing careful arguments from shoddy ones. Furthermore, people should be able to apply those same critical skills to their own observations, arguments, and conclusions, thereby becoming less bound by their own prejudices and rationalizations.

The map is organized around seven strands that reflect the ability to detect various sources of flawed arguments—*detecting bias, detecting misuse of numbers, detecting overgeneralization, detecting unfair comparisons, detecting flawed reasoning, detecting alternative explanations,* and *detecting unsupported claims*. The development of critical response skills starts with the inclination to seek evidence for claims and to question claims that are not based on fair comparisons. In middle school, the emphasis is on questioning claims and reasoning based on insufficient or flawed data. In high school, the emphasis is on detecting flaws in arguments based on faulty use of numbers and on omitting data, key assumptions, or alternative explanations from arguments.

The map draws on ideas about the importance of evidence and adequate controls on the **EVIDENCE AND REASONING IN INQUIRY** map in *Atlas 1* and on mathematical ideas on the **REASONING** map in this volume.

## NOTES

There are a fairly large number of strands in this map to reflect the wide variety of skills that students need to think critically. Throughout the map, the wording of several benchmarks has been revised to improve its clarity.

Most of the benchmarks in grades K-8 contribute to the benchmark "Notice and criticize the reasoning in arguments in which the claims are not consistent with the evidence given." At the high-school level, benchmark 12E/H4 expects students to analyze their own and other people's arguments for faulty evidence and reasoning.

## RESEARCH IN *BENCHMARKS*

Upper elementary-school students can reject a proposed experimental test where a factor whose effect is intuitively obvious is uncontrolled, at the level of "that's not fair" (Shayer & Adey, 1981). "Fairness" develops as an intuitive principle as early as 7 to 8 years of age and provides a sound basis for understanding experimental design. This intuition does not, however, develop spontaneously into a clear, generally applicable procedure for planning experiments (Wollman, 1977a, 1977b; Wollman & Lawson, 1977; Zimmerman, 2005). Although young children have a sense of what it means to run a fair test, they frequently cannot identify all the important variables, and they are more likely to control those variables that they believe will affect the result. In evaluating what can be learned from a certain experiment, students may be less likely to detect a problem in controlling variables when the outcome is expected than when the outcome is not expected (Wollman, 1977b). Student familiarity with the topic of the given experiment influences the likelihood that they will control variables (Linn & Swiney, 1981; Linn et al., 1983).

Students of all ages as well as adults may change variables one at a time to test a claim whose outcome may be construed as negative (e.g., honey makes a cake taste bad). But when the outcome is construed as positive (e.g., honey makes a cake taste good), they may hold constant what they believe is contributing to the positive outcome (Zimmerman, 2000, 2005).

After specially designed instruction, students in 8th grade are able to call attention to inadequate data resulting from lack of controls (see for example Rowell & Dawson, 1984; Ross, 1988). Explicit instruction that includes positive and negative examples of control-of-variables designs and justification for why the strategy works, combined with hands-on experimentation, can help upper elementary-school students make progress toward designing unconfounded experiments and evaluating experiments designed by others (Klahr, Chen, & Toth, 2001).

Lower elementary-school students can select conclusive (over inconclusive) tests for specific simple hypotheses (Sodian, Zaitchik, & Carey, 1991), and most 6th-graders can judge whether evidence is related to a theory, although they do not always evaluate this evidence correctly (Kuhn et al., 1988). When asked to use evidence to judge a theory, however, students of all ages may make only theory-based responses with no reference made to the presented evidence. Sometimes this appears to be because the available evidence conflicts with the students' beliefs (Kuhn et al., 1988). High-school students are more capable of evaluating theories in terms of their consistency with evidence, regardless of whether or not they believe the theory (Driver et al., 1996). This does not necessarily indicate that students appreciate the centrality of this kind of reasoning in science or that they will be inclined to evaluate claims in terms of consistency with evidence if they are not explicitly prompted (Driver et al., 1996).

Students may cite data in their arguments, but they may fail to cite sufficient evidence for claims. In addition, references to data in students' arguments often fail to articulate how specific data relate to specific claims (Sandoval & Millwood, 2005). Students may believe that data literally speak for themselves—that they are self-evident—rather than providing raw material for supporting or judging a claim (Driver et al., 1996; Sandoval & Millwood, 2005).

Some middle-school students tend to invoke personal experiences as evidence to justify a particular hypothesis. Specifically, they seem to think of evidence as selected from what is already known or from personal experience or second-hand sources, not as information produced by experiment (Roseberry et al., 1992; Ratcliffe, 1999).

Students do not necessarily consider only the evidence that is presented to them but make additional assertions about the context of the problem, or even introduce inferences that go beyond the boundaries of the evidence presented and that introduce bias in the outcome (Driver et al., 2000).

See **REASONING** for additional research.

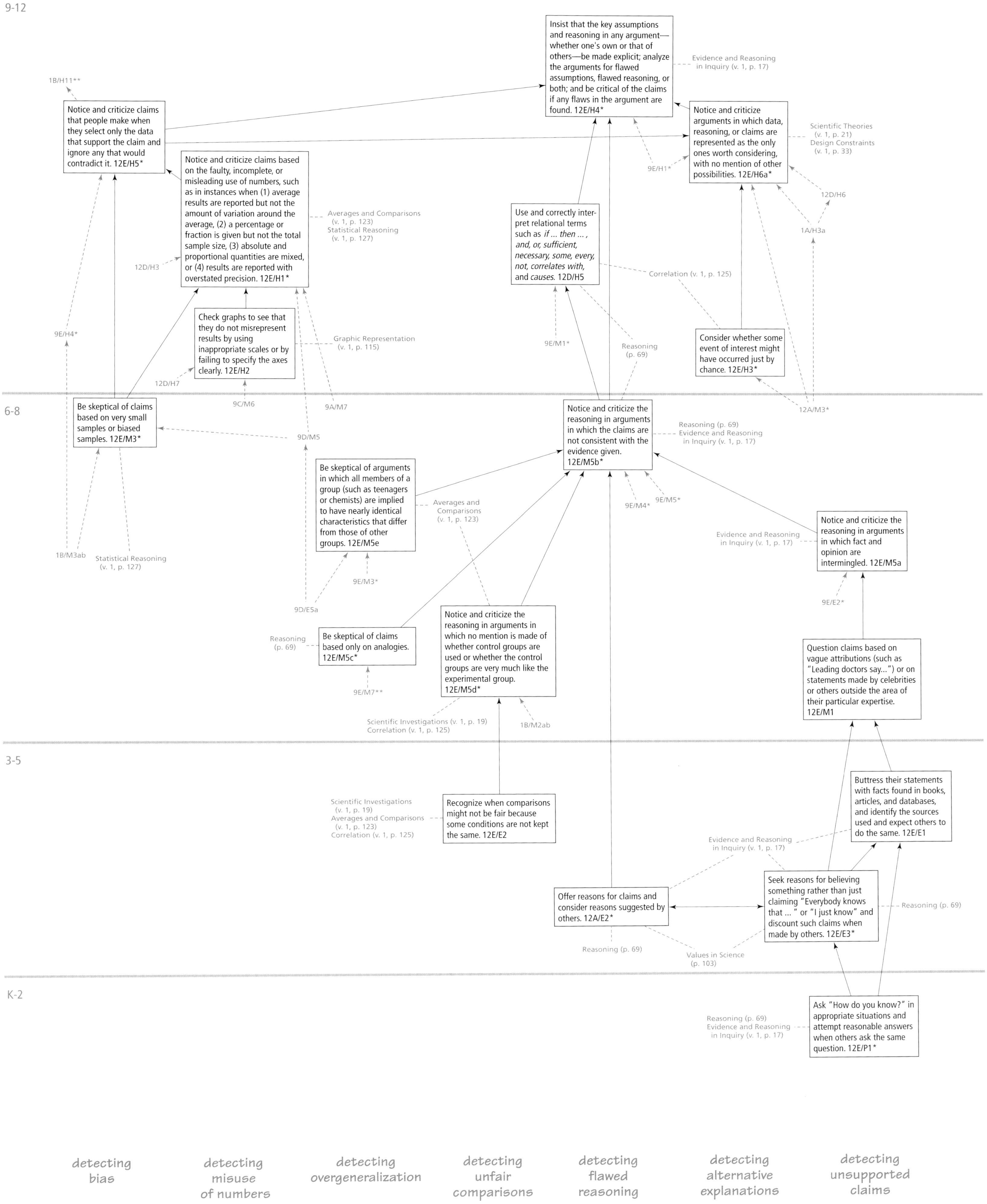
9-12
Insist that the key assumptions and reasoning in any argument—whether one's own or that of others—be made explicit; analyze the arguments for flawed assumptions, flawed reasoning, or both; and be critical of the claims if any flaws in the argument are found. 12E/H4*
Evidence and Reasoning in Inquiry (v. 1, p. 17)
1B/H11**
Notice and criticize claims that people make when they select only the data that support the claim and ignore any that would contradict it. 12E/H5*
Notice and criticize arguments in which data, reasoning, or claims are represented as the only ones worth considering, with no mention of other possibilities. 12E/H6a*
Scientific Theories (v. 1, p. 21)
Design Constraints (v. 1, p. 33)
9E/H1*
12D/H6
1A/H3a
Notice and criticize claims based on the faulty, incomplete, or misleading use of numbers, such as in instances when (1) average results are reported but not the amount of variation around the average, (2) a percentage or fraction is given but not the total sample size, (3) absolute and proportional quantities are mixed, or (4) results are reported with overstated precision. 12E/H1*
Averages and Comparisons (v. 1, p. 123)
Statistical Reasoning (v. 1, p. 127)
12D/H3
Use and correctly interpret relational terms such as if ... then ..., and, or, sufficient, necessary, some, every, not, correlates with, and causes. 12D/H5
Correlation (v. 1, p. 125)
Check graphs to see that they do not misrepresent results by using inappropriate scales or by failing to specify the axes clearly. 12E/H2
Graphic Representation (v. 1, p. 115)
9E/H4*
12D/H7
9E/M1*
Reasoning (p. 69)
Consider whether some event of interest might have occurred just by chance. 12E/H3*
6-8
9C/M6
9A/M7
12A/M3*
Be skeptical of claims based on very small samples or biased samples. 12E/M3*
9D/M5
Notice and criticize the reasoning in arguments in which the claims are not consistent with the evidence given. 12E/M5b*
Reasoning (p. 69)
Evidence and Reasoning in Inquiry (v. 1, p. 17)
Be skeptical of arguments in which all members of a group (such as teenagers or chemists) are implied to have nearly identical characteristics that differ from those of other groups. 12E/M5e
Averages and Comparisons (v. 1, p. 123)
9E/M4*
9E/M5*
Notice and criticize the reasoning in arguments in which fact and opinion are intermingled. 12E/M5a
Evidence and Reasoning in Inquiry (v. 1, p. 17)
1B/M3ab
Statistical Reasoning (v. 1, p. 127)
9E/M3*
9D/E5a
9E/E2*
Reasoning (p. 69)
Be skeptical of claims based only on analogies. 12E/M5c*
Notice and criticize the reasoning in arguments in which no mention is made of whether control groups are used or whether the control groups are very much like the experimental group. 12E/M5d*
Question claims based on vague attributions (such as "Leading doctors say...") or on statements made by celebrities or others outside the area of their particular expertise. 12E/M1
9E/M7**
Scientific Investigations (v. 1, p. 19)
Correlation (v. 1, p. 125)
1B/M2ab
3-5
Scientific Investigations (v. 1, p. 19)
Averages and Comparisons (v. 1, p. 123)
Correlation (v. 1, p. 125)
Recognize when comparisons might not be fair because some conditions are not kept the same. 12E/E2
Buttress their statements with facts found in books, articles, and databases, and identify the sources used and expect others to do the same. 12E/E1
Evidence and Reasoning in Inquiry (v. 1, p. 17)
Offer reasons for claims and consider reasons suggested by others. 12A/E2*
Seek reasons for believing something rather than just claiming "Everybody knows that ... " or "I just know" and discount such claims when made by others. 12E/E3*
Reasoning (p. 69)
Reasoning (p. 69)
Values in Science (p. 103)
K-2
Reasoning (p. 69)
Evidence and Reasoning in Inquiry (v. 1, p. 17)
Ask "How do you know?" in appropriate situations and attempt reasonable answers when others ask the same question. 12E/P1*
detecting bias
detecting misuse of numbers
detecting overgeneralization
detecting unfair comparisons
detecting flawed reasoning
detecting alternative explanations
detecting unsupported claims

Paul Klee *Mural from the Temple of Longing, Thither,* 1922/30

# REFLECTIONS

*Atlas of Science Literacy* provides a set of strand maps that show connections among the benchmarks that lead to literacy in science, mathematics, and technology. It also promotes thinking through how students' understanding develops over time in order to attain science literacy. In this section we reflect on the strand maps—their history, issues raised by the mapping process, and ways in which the strand-mapping process can be used by others.

## ORIGIN OF STRAND MAPS

The strand maps grew out of a conceptual perspective on K-12 education in science, mathematics, and technology that predates the publication of *Science for All Americans* or *Benchmarks for Science Literacy*. The kind of thinking that goes into mapping is inextricably tied to Project 2061's vision of science literacy and of how science education needs to change.

### CREATING *BENCHMARKS FOR SCIENCE LITERACY*

From the beginning, Project 2061 recognized the need to think through the conceptual steps along the way to adult science literacy. Deciding on adult literacy goals in *Science for All Americans* often involved imagining what the conceptual

Figure 6: An early map on evolution.

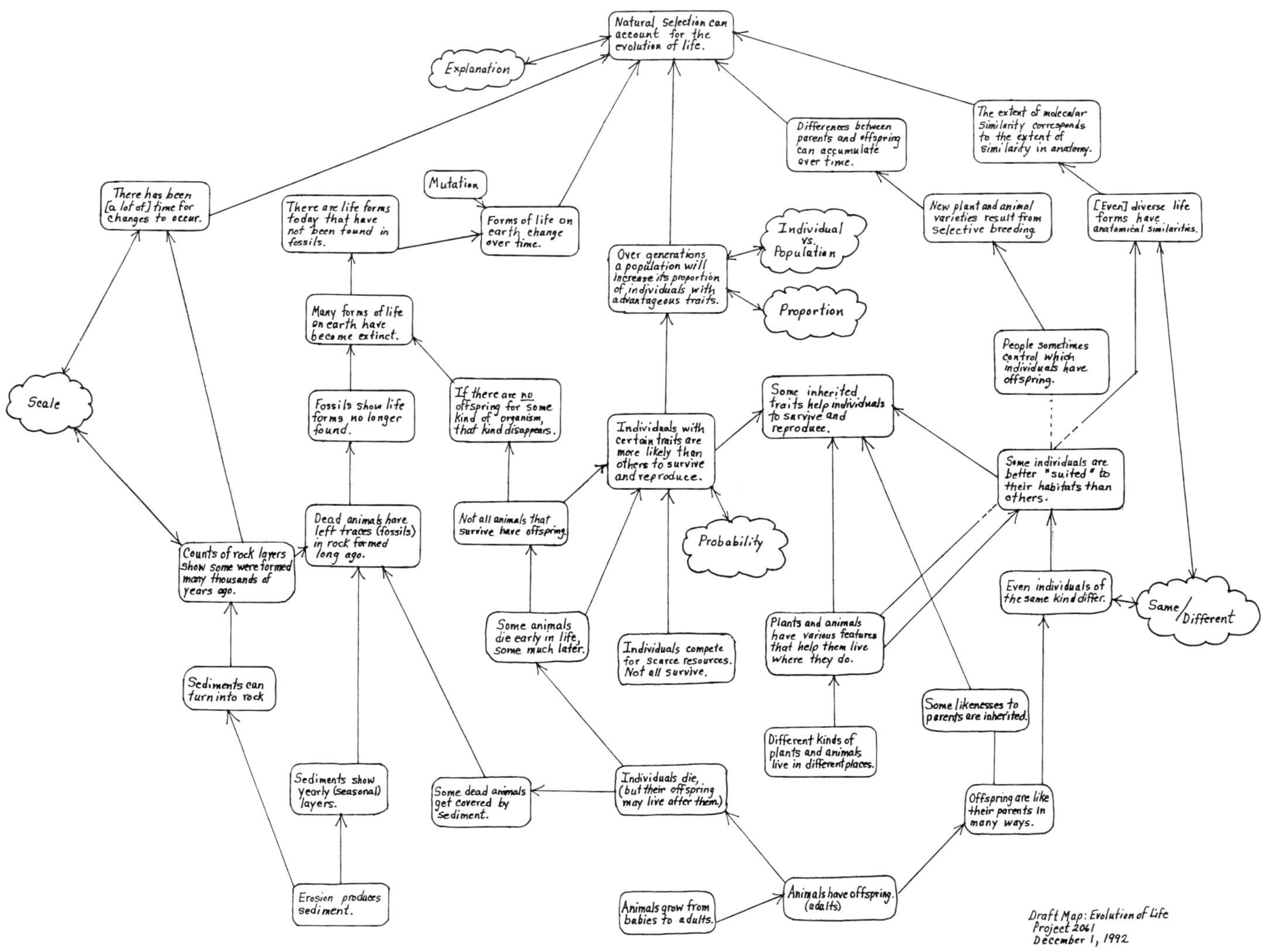

steps would be and whether they were plausible for all students. After publishing *Science for All Americans* in 1989, Project 2061 engaged six school-district teams in building curriculum models aimed at the literacy goals laid out in *Science for All Americans*. The Project 2061 staff prepared this message to convey to the teams its commitment to students' growth of understanding:

> If we invest our energies in selecting or inventing activities and placing them intuitively at different grade levels, we will fall short of the quality of innovation that Project 2061 intends. The job is rather to think through the entire flow of learning, including the major connections among ideas, so as to identify the kinds of learning experiences that would optimally contribute to students growing along those lines.

Figure 7: A partial strand about the water cycle.

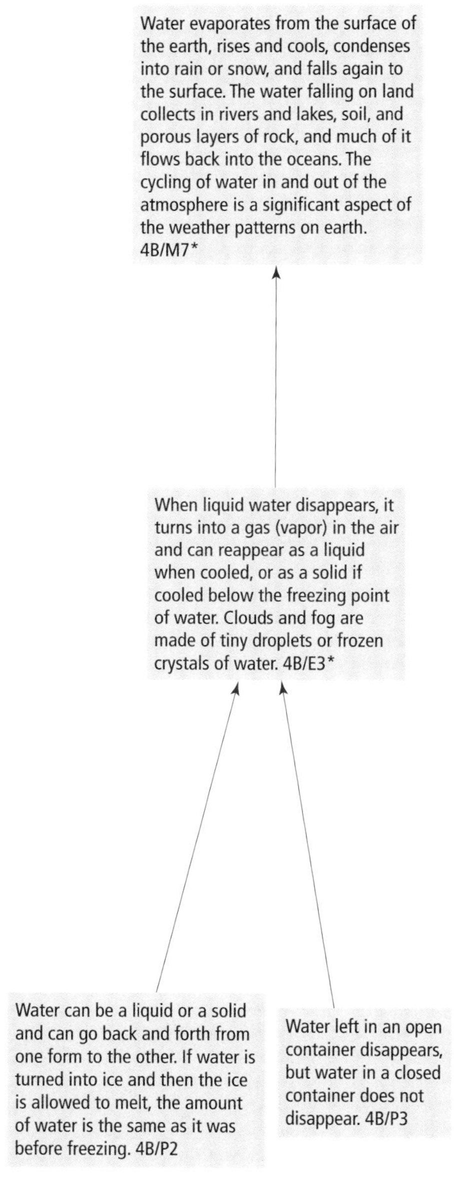


Before beginning to design curriculum models, each team began "mapping" the development of ideas and skills from kindergarten through high-school graduation. The teams soon agreed on the need for a common set of precursors to the literacy goals in *Science for All Americans* that could be the basis for designing and organizing learning experiences.

The effort to specify a common set of precursors led to *Benchmarks for Science Literacy*. No complete set of maps was sketched in the development of *Benchmarks*, but attention to connections among the benchmarks was a major part of the effort. A map about evolution that was used in the development of *Benchmarks* is shown in Figure 6.

An early draft of *Benchmarks* included superscript letters placed next to each benchmark to indicate strands of development across grades. This coding reflected the kind of thinking behind *Benchmarks*, but reviewers agreed the superscripts were difficult to understand, and they were not included in the published version. But the importance of thinking about connections among benchmarks was not forgotten.

### MAKING CONNECTIONS EXPLICIT

As educators began using *Benchmarks*, it became clear that the detailed connections among the benchmarks, if made explicit and formalized in maps, could help people understand and use *Benchmarks* better.

As carefully as *Benchmarks* was written, it still requires considerable study to see what specific ideas and skills are expressed in the benchmarks and how they relate to each other. In Project 2061 workshops, for example, participants study the precise meaning of individual benchmarks by reading corresponding sections in *Science for All Americans* and in the *Benchmarks* essays, and by studying benchmarks at different grade levels, the "Also See" boxes, the cognitive research, and a relevant map. Maps help clarify the context of a benchmark, showing where it comes from and where it leads, and how it relates to other benchmarks.

Consider the following primary-school benchmark:

> Water left in an open container disappears, but water in a closed container does not disappear.

When workshop participants first consider this benchmark on its own, they often assume it asks students to understand evaporation. But by studying the benchmark and its connections to others on a map, they can see that the benchmark leads to a later benchmark about water turning into vapor in the air. (The growth of understanding about evaporation is represented in part of a map shown in Figure 7.) This context makes it clear that primary-school students need only know that water in an open container disappears—and that they can wait until grades 3-5 to know where it goes and until grades 6-8 to know the mechanism, evaporation.

*Benchmarks* users also sought guidance in identifying important connections to be made among topics and ideas within a grade range. For example, which ideas about interdependence of life are particularly relevant to ideas about the use of Earth's resources? Which connections between ideas about energy transformations and ideas about the flow of energy in ecosystems are important to note? When should ideas about atoms and molecules be connected to ideas about the interdependence of life? Which specific ideas from science, mathematics and technology could be helpful in understanding general ideas about models, constancy, and patterns of change? Which ideas from the history of science episodes best illustrate ideas about the nature of science? The "Also See" boxes in *Benchmarks* (Figure 8) had pointed users to other relevant chapters but not to other relevant sections within the same chapter and, moreover, had not made the connections between particular benchmarks explicit. Maps, with their numerous and varied types of cross-references, can serve to make these connections explicit.

## CREATING *ATLAS OF SCIENCE LITERACY*

The development of *Atlas* drew on the work done on maps for other Project 2061 publications and reflected the same spirit of thinking through growth of understanding. The additional detail provided in *Atlas* maps required more involved analysis of each benchmark, including the dissection of benchmarks into key ideas and the identification of connections among those key ideas and between them and ideas in other benchmarks. Sometimes the analysis revealed the need for intermediate steps between pairs of ideas, sometimes it identified a missing culminating idea to tie two or more ideas together, and occasionally it suggested the need for updating the science itself. As with all Project 2061 publications, the *Atlas* maps were extensively reviewed by scientists and educators, both individually and in groups. Maps published in Volume 2 have also benefited from suggestions made over the past six years by participants in Project 2061's *Atlas* workshops, by curriculum materials developers, and by our own attempts to further clarify particular benchmarks for assessment analysis and development.

Because of the perspective that mapping provides on the coherence and completeness of a set of learning goals, the *Atlas* maps include some benchmarks that are modifications of, or additions to, those presented in *Benchmarks*. As noted in the description of strand maps in the Introduction, new ideas are drawn mainly from *Science for All Americans*, from essays in *Benchmarks for Science Literacy*, and from the *National Science Education Standards*; hence these ideas are already part of the national consensus on science literacy.

Figure 8: The "Also See" box in *Benchmarks* points users to related ideas in other chapters.

ALSO SEE

| | | |
|---|---|---|
| *Chapter* 1 | C | THE SCIENTIFIC ENTERPRISE (ethics, effects on environment) |
| 3 | C | ISSUES IN TECHNOLOGY (impact of technology on ecosystem) |
| 4 | B | THE EARTH (habitat, needed resources) |
| | C | PROCESSES THAT SHAPE THE EARTH (climate and habitat) |
| 8 | A | AGRICULTURE (ecosystems) |
| | B | MATERIALS AND MANUFACTURING (loss and recovery of the earth's resources) |
| 11 | A | SYSTEMS |
| | B | MODELS (modeling ecosystems) |
| | C | CONSTANCY AND CHANGE (ecosystems) |

transfer of thermal energy through materials (4E/M3). The same map also illustrates a looser kind of connection: The idea in primary grades about the ability to keep track of things (11C/P2) serves as a helpful precursor to the middle-school idea that whenever energy appears in one place it must have disappeared from another (4E/M1). Other benchmarks may simply support or enhance the understanding of a later benchmark.

Connections can also differ in the kind of relationship they imply. An arrow may mean that the earlier benchmark provides some fact that serves as evidence to support an idea in the later benchmark. In the **DIVERSITY OF LIFE** map, for example, a middle-school benchmark stating that patterns of development are similar in humans and other vertebrates serves as evidence to support the high-school benchmark that points out that similar patterns of development and internal anatomy suggest relatedness among organisms.

An arrow may also mean that an earlier benchmark provides an example or instance of a general principle or generalization in a later benchmark. In the **PATTERNS OF CHANGE** map, the K-2 benchmarks (shown as off-map connections) provide examples—daily pattern of change in the appearance of the sun, monthly pattern of change in the appearance of the moon, yearly pattern of change in temperature—of the generalization in a grades 3-5 benchmark that some things in nature have a repeating pattern. At the same time, understanding the generalization can often reinforce the earlier ideas that lead to it.

These different aspects of the connections represented by the arrows are not made explicit in the maps; they are often too complex to represent effectively. But the reader should keep such considerations in mind while studying the maps. They may have important implications for designing curriculum, instruction, or assessments.

## GENERALIZATIONS AND EXAMPLES

The maps include both general principles and specific examples of those principles, at all levels of sophistication. As is the case with ideas about patterns of change mentioned above, students usually need to have experience with several relevant examples, and a generalization may also shed light back on the specific examples. Indeed, reinforcement of earlier knowledge is a likely feature of all learning.

Practically, there is a significant amount of back-and-forth between specific examples and emerging generalizations. Sometimes, a simple version of a general principle can make sense early on (though it certainly should be strengthened and extended later). And some specific examples must wait until later because of their level of sophistication or because they depend on some other knowledge not yet developed. Consider the following primary-school benchmark on the **ENERGY TRANSFORMATIONS** map:

> People can keep track of things, seeing where they come from and where they go.

With adequate examples, this simple generalization can have some meaning for primary-school students. It then provides a foundation for more specific examples later, such as the concept of energy conservation, which, in turn, contributes to more sophisticated general notions (such as principles of conservation, equilibrium, etc). General principles and examples reinforce one another as they become more sophisticated; no absolute rule about which should come first is adequate to characterize growth of understanding.

To provide guidance to developers of curriculum and instructional materials, and to foster efficient use of students' time, *Atlas* maps present specific examples as precursors to general principles whenever specific examples exist in *Benchmarks*. When specific examples do not exist, it is mainly because we did not think that any particular example should be taught or tested and hence should not be included in *Benchmarks*. As an illustration, consider the following middle-school benchmark about systems that appears on the **BASIC FUNCTIONS** map:

> Thinking about things as systems means looking for how every part relates to others. The output from one part of a system (which can include material, energy, or information) can become the input to other parts. Such feedback can serve to control what goes on in the system as a whole.

Its only precursor on the map is another general principle, which indicates that when things are made of parts, the parts may influence each other. The lack of specific precursors does not mean that the systems benchmark should be taught in the abstract, but rather that curriculum developers and teachers can choose any appropriate examples they like, such as examples dealing with designed systems, which are often more transparent to students. The generalization about systems, in turn, contributes to the benchmarks on body systems above it on the map; these benchmarks both benefit from and help to refine students' understanding of systems.

### MAP SIZE

Connections among the benchmarks are prolific enough so that all the benchmarks could conceivably be put together into one vast map. But design constraints, readability, and usefulness demanded that some reasonable number of related benchmarks be considered a single map. Decisions were made according to the content of the map, not according to a rigid rule. A quick scan of *Atlas 1* and *Atlas 2* will reveal that not all maps are of the same size.

Sometimes there are "natural" boundaries for a map—where one set of benchmarks seems more cohesive than others. Compromises had to be made between natural coherence, completeness, and readability. The inclusion of specific off-map connections in *Atlas 2* is intended to serve the needs of large-scale curriculum planning, while also making it possible to create maps that are conceptually coherent and scaled to a reasonable page size.

## CONSTRUCTION OF STRAND MAPS

Thinking through the growth of understanding that leads to literacy in science, mathematics, and technology is a demanding and time-consuming task. By displaying the relevant benchmarks and connections among them, *Atlas* can save educators an immense time investment. We hope, nonetheless, that teachers and other educators will tinker with the maps presented here or use them as an inspiration for making new maps, such as on topics outside of science, mathematics, or technology or on topics that go beyond science literacy. This section describes the construction of maps for two reasons: to enhance the reader's understanding of the maps and to offer suggestions on how to construct their own maps. For all of the tasks in which studying a map can be useful, having firsthand experience in constructing the map can often provide deeper insights into those tasks. Furthermore, map making is perhaps the most effective way to gain an understanding of any particular set of learning goals and the content they include.

Like any complex process, mapping cannot be adequately characterized by a linear sequence of steps, such as the one described here. Some elements of map construction mentioned later will inform earlier ones, and the actual process involves a lot of moving back and forth between different steps. Nevertheless, what follows provides a useful outline of the most important aspects of the mapping process.

### IDENTIFYING A SET OF LEARNING GOALS

Mapping can be attempted with any reasonably coherent and specific set of learning goals, such as those contained in *Benchmarks for Science Literacy*, in the *National Science Education Standards*, or in state and local curriculum frameworks or standards. It is also possible to start with a set of learning goals established for a single course.

To lend themselves to mapping, however, learning goals have to be specific enough for clear connections to be made between them. A learning goal that states students should understand "energy" could be connected to a huge number of benchmarks or other learning goals. But without specifying exactly what it is about energy that students need to understand and at what level of sophistication, it is difficult to make connections that clarify students' growth of understanding—that is, what they would need to know beforehand to prepare and what later ideas their understanding of energy could contribute to.

To produce thoughtful maps, a set of learning goals also has to be complete and coherent enough to provide a suitable framework—all the conceptually important ideas or skills need to be there in the first place. Of course, the process of mapping is also a good way to determine whether a set of goals is complete and coherent, and what ideas or skills need to be added if it is not.

## SELECTING BENCHMARKS

Making a map begins with the selection of a particular topic that will be the focus of the map and the identification of relevant learning goals at each grade range. In working with *Benchmarks*, many of the benchmarks for any particular topic are together in one section, but typically there are relevant benchmarks in other sections as well. For example, the **MATERIALS SCIENCE** map has 14 benchmarks from the Materials and Manufacturing section of Chapter 8: THE DESIGNED WORLD. But it also has 11 benchmarks from 3 different sections of Chapter 4: THE PHYSICAL SETTING. In finding other benchmarks, the "Also See" boxes that appear in every section of *Benchmarks for Science Literacy* are useful, but it is also helpful to review benchmarks in relevant sections of the same chapter, since these are not referenced in the "Also See" boxes.

The selection of benchmarks is also influenced by careful consideration of what the essential elements of the topic are and by how those elements will be represented by the separate strands on the map. A careful review of the appropriate sections of *Science for All Americans* provides critical guidance for this task. The index entry for each section of *Science for All Americans* includes a list of the topics in that section, and this list was frequently used in coming up with titles for the strands on the *Atlas* maps.

For those working with other sets of science, mathematics, or technology learning goals, referring to the cognitive research summaries opposite each map can take some of the guesswork out of positioning those learning goals in grade levels.

## MAKING CONNECTIONS

The very nature of the maps leads to an iterative development process. As a starting point, it helps to place the relevant benchmarks on a single page arranged in their respective grade ranges. Individual benchmarks are analyzed for their meaning and tentatively assigned to a strand. If a benchmark is seen to include multiple key ideas, then the key ideas may be separated into different boxes to increase the precision of meaning of connections. A connection is drawn whenever achieving one idea seems likely to contribute to achieving another. This involves deciding not only whether two ideas are conceptually related, but also what the relationship implies about which of the ideas students are likely to need and be able to achieve first. Although maps do not prescribe instruction, deciding on what connects to what involves imagining possible ways students might attain the benchmarks.

There may be several reasons not to make an explicit connection between related benchmarks. Another benchmark may be needed as an intermediate step between them. Sometimes the relationship between benchmarks is best expressed by their mutual contribution to a third benchmark such as when a third benchmark explains observations described in the two other benchmarks. And sometimes the relationship between benchmarks makes sense to someone with a sophisticated grasp of the topic but is not helpful in terms of students' growth of understanding. In this case, the connection between two benchmarks is only apparent once one understands the theory that explains them both.

As benchmarks are linked, strand assignments are revisited. Strands are sometimes combined together or divided in two depending on the number of benchmarks they contain and how the layout of the map evolves. *Science for All Americans*, the *National Science Education Standards*, and research on students' understanding of the topic are consulted to look for any important ideas—such as bridging or culminating ideas—that should be added to the map.

Where should a map stop? For *Atlas* maps, the top and bottom are easy—kindergarten and 12$^{th}$ grade. But the "sides" are more difficult. Eventually, the mapmaker must decide how many benchmarks need to be viewed at the same time and how many can fit on the page. Ultimately, each map should represent enough benchmarks and connections to offer insight and perspective on a topic, but not so many that the map becomes unapproachable or incomprehensible. As the map takes shape, story lines come into focus, fostering clarification at a finer grain size. Some benchmarks, and in some cases entire strands, are removed from the map to sharpen the focus. The use of off-map connections in *Atlas 2* allows maps to convey some of the relationships among ideas that are not displayed in their full text on the page.

The mapmaking process also provides an opportunity to look carefully at the wording of individual benchmarks. Many benchmarks in this volume have been edited as a result of this close scrutiny. Sometimes it is simply a small change in wording to make the meaning of the benchmark clearer. In other cases, the changes have been made to better reflect the original intent of *Science for All Americans*.

Seeking feedback on maps from others is an essential step in the process. When possible, Project 2061 consults with educators who are developing curriculum materials or assessments on topics that are being mapped; often these educators are dealing with many of the same issues that are faced in making a map. In addition, every map is formally reviewed by scientists and by educators with knowledge of the relevant content and learning research.

Maps often go through more than a dozen iterations during the development process. Because of the close interconnections among the maps, making changes on one map often requires rethinking the work done on other maps.

## THE FUTURE OF STRAND MAPS

The development of maps for *Atlas 2*, along with Project 2061's assessment research and development, has led to the review of many maps in *Atlas 1*. Eventually, all of the maps in the first volume will be updated to reflect our latest thinking and research findings. In addition, the wording of individual benchmarks and the topography of the maps themselves will be reviewed in light of the work that has been done for *Atlas 2*. And, as always, new scientific understandings and new research on student learning will lead to modifications of the maps. A planned future edition of *Atlas* will bring the two volumes together. In the meantime, we invite educators to use the maps in their work and share their ideas for improving it.

# INDEX OF MAPPED BENCHMARKS

This index will best serve the reader who is working with a particular benchmark and wants to find the maps in both Volume 1 and Volume 2 of *Atlas of Science Literacy* in which that benchmark appears. It is not an index of terms or topics in the benchmarks or in this book. The index is organized according to the structure of the chapters and sections in *Benchmarks for Science Literacy*. Each benchmark listed in the index is identified with a unique code that can also be found at the end of that benchmark wherever it appears on a map. The code indicates the chapter, section, and grade range of the map in question (see p. *xxii* of the Introduction for a full explanation of the codes). Each index entry provides the names of the maps, along with the volume and page numbers, on which the benchmark can be found.

## 1 THE NATURE OF SCIENCE

### A THE SCIENTIFIC WORLD VIEW



### B SCIENTIFIC INQUIRY

### C THE SCIENTIFIC ENTERPRISE



## 2 THE NATURE OF MATHEMATICS

### A PATTERNS AND RELATIONSHIPS



### B MATHEMATICS, SCIENCE, AND TECHNOLOGY



### C MATHEMATICAL INQUIRY

## 3 THE NATURE OF TECHNOLOGY

### A TECHNOLOGY AND SCIENCE



### B DESIGN AND SYSTEMS



### C ISSUES IN TECHNOLOGY



## 4 THE PHYSICAL SETTING

### A THE UNIVERSE

**B THE EARTH**



**C PROCESSES THAT SHAPE THE EARTH**

### F MOTION



### G FORCES OF NATURE



## 5 THE LIVING ENVIRONMENT

### A DIVERSITY OF LIFE

## E FLOW OF MATTER AND ENERGY



## F EVOLUTION OF LIFE



# 6 THE HUMAN ORGANISM

## A HUMAN IDENTITY



## B HUMAN DEVELOPMENT

## C BASIC FUNCTIONS



## D LEARNING



## E PHYSICAL HEALTH



## F MENTAL HEALTH

### F SOCIAL CONFLICT



### G GLOBAL INTERDEPENDENCE



## 8 THE DESIGNED WORLD

### A AGRICULTURE



### B MATERIALS AND MANUFACTURING

## 9 THE MATHEMATICAL WORLD

## B SYMBOLIC RELATIONSHIPS



## C SHAPES



## D UNCERTAINTY

## 10 HISTORICAL PERSPECTIVES

## 11 COMMON THEMES

### D SCALE



## 12 HABITS OF MIND

### A VALUES AND ATTITUDES



### B COMPUTATION AND ESTIMATION

## C MANIPULATION AND OBSERVATION



## D COMMUNICATION SKILLS



## E CRITICAL-RESPONSE SKILLS

# REFERENCES

The following includes references to the research cited in the map commentaries and to other resources that were used in developing *Atlas 2*. Most of the research references are taken from Chapter 15: THE RESEARCH BASE of *Benchmarks for Science Literacy*.


## A

Aikenhead, G. S. (1987). High school graduates' beliefs about science-technology-society III. Characteristics and limitations of scientific knowledge. *Science Education*, 71, 459-487.

American Association for the Advancement of Science. (1989). *Science for all Americans*. New York: Oxford University Press.

American Association for the Advancement of Science. (1993). *Benchmarks for science literacy*. New York: Oxford University Press.

American Association for the Advancement of Science. (2001). *Atlas of science literacy*. Washington, DC: Author.

Ametller, J., & Pinto, R. (2002). Students' reading of innovative images of energy at secondary school level. *International Journal of Science Education*, 24, 285-312.

Amsel, E., & Brock, S. (1996). The development of evidence evaluation skills. *Cognitive Development*, 11, 523-550.

Andersson, B., & Wallin, A. (2000). Students' understanding of the greenhouse effect, the societal consequences of reducing $CO_2$ emissions and the problem of ozone layer depletion. *Journal of Research in Science Teaching*, 37, 1096-1111.

Arnaudin, M. W., & Mintzes, J. J. (1985). Students' alternative conceptions of the human circulatory system: A cross age study. *Science Education*, 69, 721-733.

Arnaudin, M. W., & Mintzes, J. J. (1986). The cardiovascular system: Children's conceptions and misconceptions. *Science and Children*, 23(5), 48-51.

Arnold, M., & Millar, R. (1987). Being constructive: An alternative approach to the teaching of introductory ideas in electricity. *International Journal of Science Education*, 9, 553-563.

Arons, A. (1997). *Teaching introductory physics*. New York: John Wiley & Sons, Inc.

Atwood, R., & Atwood, V. (1996). Preservice elementary teachers' conceptions of the causes of seasons. *Journal of Research in Science Teaching*, 33, 553-563.

## B

Bar, V. (1989). Children's views about the water cycle. *Science Education*, 73, 481-500.

Behr, M., Lesh, R., Post, T., & Silver, E. (1983). Rational number concepts. In R. Lesh & M. Lindau (Eds.), *Acquisition of mathematical concepts and processes* (pp. 91-126). New York: Academic Press.

Bell, A., Fischbein, E., & Greer, B. (1984). Choice of operation in verbal arithmetic problems: The effects of number size, problem, structure, and context. *Educational Studies in Mathematics*, 15, 129-147.

Bell, A., Swan, M., & Taylor, G. (1981). Choice of operations in verbal problems with decimal numbers. *Educational Studies in Mathematics*, 12, 399-420.

Bell, B. (1981). When is an animal, not an animal? *Journal of Biological Education*, 15, 213-218.

Bell, B., & Freyberg, P. (1985). Language in the science classroom. In R. Osborne & P. Freyberg (Eds.), *Learning in science* (pp. 29-40). Auckland, NZ: Heinemann.

Benander, L., & Clement, J. (1985). *Catalogue of error patterns observed in courses in basic mathematics*. Unpublished manuscript. (ERIC Reproduction Service No. ED 287 672).

Bernstein, A. C., & Cowan, P. A. (1975). Children's concepts of how people get babies. *Child Development*, 46, 77-91.

Berti, A. E. (1988). The development of political understanding in children between 6-15 years old. *Human Relations*, 41, 437-446.

Berti, A. E. (1992). Acquisition of the profit concept by third-grade children. *Contemporary Educational Psychology*, 17, 293-299.

Berti, A. E., & Bombi, A. S. (1988). *The child's construction of economics*. Cambridge: Cambridge University Press.

Berti, A. E., & Grivet, A. (1990). The development of economic reasoning in children from 8 to 13 years old: Price mechanism. *Contributi di Psicologia*, 3(III), 37-47.

Black, P. (1990). Can pupils design their own experiments? In *Proceedings of the international conference on physics education through experiments* (pp. 281-299). Tianjin, China: ICPE.

Blum, L. H. (1977). Health information via mass media: Study of the individual's concepts of the body and its parts. *Psychological Reports*, 40, 991-999.

Borges, A., & Gilbert, J. (1999). Mental models of electricity. *International Journal of Science Education*, 21, 95-117.

Boylan, C., Hill, D., Wallace, A., & Wheeler, A. (1992). Beyond stereotypes. *Science Education*, 76, 465-476.

Brook, A., & Driver, R. (1984). *Aspects of secondary students' understanding of energy: Summary report*. Leeds, UK: University of Leeds, Centre for Studies in Science and Mathematics Education.

Brook, A., & Driver, R. (1986). *The construction of meaning and conceptual change in the classroom: Case studies on energy*. Leeds, UK: University of Leeds, Centre for Studies in Science and Mathematics Education.

Brook, A., & Wells, P. (1988). Conserving the circus: An alternative approach to teaching and learning about energy. *Physics Education*, 23, 80-85.

Brown, C., Carpenter, T., Kouba, V., Lindquist, M., Silver, E., & Swafford, J. (May 1988). Secondary school results from the fourth NAEP mathematics assessment: Algebra, geometry, mathematical methods, and attitudes. *Mathematics Teacher*, 337-347, 397.

Brown, J., & VanLehn, K. (1982). Towards a generative theory of "bugs." In T. P. Carpenter, J. Moser, & T. Romberg (Eds.), *Addition and subtraction: A cognitive perspective* (pp. 117-135). Hillsdale, NJ: Lawrence Erlbaum Associates.

Brumby, M. (1982). Students' perceptions of the concept of life. *Science Education*, 66, 613-622.

Burger, W., & Shaughnessy, J. (1986). Characterizing the van Hiele levels of development in geometry. *Journal for Research in Mathematics Education*, 17, 31-48.

## C

Carey, S. (1985). *Conceptual change in childhood*. Cambridge, MA: MIT Press.

Carey, S., Evans, R., Honda, M., Jay, E., & Unger, C. (1989). An experiment is when you try it and see if it works: A study of grade 7 students' understanding of the construction of scientific knowledge. *International Journal of Science Education*, 11, 514-529.

Carpenter, T., Corbitt, M., Kepner, H., Lindquist, M., & Reys, R. (1981). Decimals: Results and implications from the second NAEP mathematics assessment. *Arithmetic Teacher*, 28(8), 34-37.

Carpenter, T. P., Franke, M. L., & Levi, L. (2003). *Thinking mathematically: Integrating arithmetic and algebra in the elementary school*. Portsmouth, NH: Heinemann.

Carpenter, T., & Moser, J. (1983). The acquisition of addition and subtraction concepts. In R. Lesh & M. Landau (Eds.), *Acquisition of mathematics: Concepts and processes* (pp. 7-44). New York: Academic Press.

Carr, M., & Kirkwood, V. (1988). Teaching and learning about energy in New Zealand secondary school junior science classrooms. *Physics Education*, 23, 86-91.

Case, R., & Sowder, J. (1990). The development of computational estimation: A neo-Piagetian analysis. *Cognition and Instruction*, 7, 79-104.

Chinn, C., & Malhotra, B. A. (2002). Children's responses to anomalous scientific data: How is conceptual change impeded? *Journal of Educational Psychology*, 19, 327-343.

Clark, D. (2006). Longitudinal conceptual change in students' understanding of thermal equilibrium: An examination of the process of conceptual restructuring. *Cognition and Instruction*, 24, 467-563.

Clark, D., & Jorde, D. (2004). Helping students revise disruptive experientially-supported ideas about thermodynamics: Computer visualizations and tactile models. *Journal of Research in Science Teaching*, 41, 1-23.

Clark, D., & Linn, M. (2003). Scaffolding knowledge integration through curricular depth. *Journal of Learning Sciences*, 12, 451-494.

Clements, D., & Battista, M. (1989). Learning of geometric concepts in a Logo environment. *Journal for Research in Mathematics Education*, 20, 450 467.

Clements, D., & Battista, M. (1990). The effects of Logo on children's conceptualizations of angle and polygons. *Journal for Research in Mathematics Education*, 21, 356-371.

Clements, D., & Battista, M. (1992). Geometry and spatial reasoning. In D. Grouws (Ed.), *Handbook of research on mathematics teaching and learning* (pp. 420-464). New York: Macmillan Publishing Company.

Closset, J. L. (1983). Sequential reasoning in electricity. In *Research in physics education: Proceedings of the first international workshop* (pp. 313-319). Paris: Centre National de la Recherche Scientifique.

Clough, E. E., & Wood-Robinson, C. (1985). How secondary students interpret instances of biological adaptation. *Journal of Biological Education,* 19, 125-130.

Coll, R., France, B., & Taylor, I. (2005). The role of models/ and analogies in science education: Implications from research. *International Journal of Science Education,* 27, 183-198.

Connell, R. (1971). *The child's construction of politics.* Calton, Australia: Melbourne University Press.

Contento, I. (1981). Children's thinking about food and eating: A Piagetian-based study. *Journal of Nutrition Education,* 13(1), 86-90.

## D

Damon, W. (1977). *The social world of the child.* San Francisco: Jossey-Bass.

Donaldson, M. (1978). *Children's minds.* New York: W. W. Norton & Company.

Dove, J. (1998). Alternative conceptions about the weather. *School Science Review,* 79, 65-69.

Downey, M. & Levstik, L. (1988, September). Teaching and learning history:The research base. *Social Education,* 336-342.

Driver, R., Leach, J., Millar, R., & Scott, P. (1996). *Young people's images of science.* Buckingham: Open University Press.

Driver, R., Newton, P., & Osborne, J. (2000). Establishing the norms of scientific argumentation in classrooms. *Science Education,* 84, 287-312

Driver, R., Squires, A., Rushworth, P., & Wood-Robinson, V. (1994a). *Making sense of secondary science: Research into children's ideas.* New York: Routledge.

Driver, R., Squires, A., Rushworth, P., & Wood-Robinson, V. (1994b). *Making sense of secondary science: Support materials for teachers.* New York: Routledge.

Duit, R., & Hauessler, P. (1994). Learning and teaching energy. In P. Fensham, R. Gunstone, & R. White (Eds.), *The content of science* (pp. 185-200). London: The Falmer Press.

## E

Edwards, C. P. (1986). *Promoting social and moral development in young children.* New York: Teachers College Press.

Egan, K. (1982, March). Teaching history to young children. *Phi Delta Kappan,* 439-441.

Ericksen, D. (1991). Students ability to recognize patterns. *School Science and Mathematics,* 91, 255-258.

Erickson, G. (1985). Heat and temperature: An overview of pupils' ideas. In R. Driver, E. Guesne, & A. Tiberghien (Eds.), *Children's ideas in science* (pp. 55-66). Milton Keynes, UK: Open University Press.

Eylon, B. & Ganiel, U. (1990). Macro-micro relationships: The missing link between electrostatics and electrodynamics in students' reasoning. *International Journal of Science Education,* 12, 79-94.

## F

Fleming, R. (1986a). Adolescent reasoning in socio-scientific issues. Part I: Social cognition. *Journal of Research in Science Teaching,* 23, 677-687.

Fleming, R. (1986b). Adolescent reasoning in socio-scientific issues. Part II: Nonsocial cognition. *Journal of Research in Science Teaching,* 23, 688-698.

Fleming, R. (1987). High school graduates' beliefs about science-technology-society II. The interaction among science, technology, society. *Science Education,* 71, 163-186.

Fort, D., & Varney, H. (1989). How students see scientists: Mostly male, mostly white, and mostly benevolent. *Science and Children,* 26(8), 8-13.

Fuson, K. (1988). *Children's counting and concepts of number.* New York: Springer Verlag.

Fuson, K. (1992). Research on whole number addition and subtraction. In D. A. Grouws (Ed.), *Handbook of research on mathematics teaching and learning* (pp. 243-275). New York: Macmillan Publishing Company.

Fuson, K., Richards, J., & Briars, D. (1982). The acquisition and elaboration of the number word sequence. In C. Brainerd (Ed.), *Progress in cognitive development research Vol. 1: Children's logical and mathematical cognition* (pp. 33-92). New York: Springer Verlag.

Fuson, K., & Willis, G. (1989). Second graders' use of schematic drawings in solving addition and subtraction word problems. *Journal of Educational Psychology,* 81, 514-520.

## G

Galili, I., & Lavrik, V. (1998). Flux concept in learning about light: A critique of the present situation. *International Journal of Science Education,* 82, 591-613.

Gauld, C. (1989). A study of pupils' responses to empirical evidence. In R. Millar (Ed.), Doing science: *Images of science in science education* (pp. 62-82). London: Falmer Press.

Gega, P. (1986). *Science in elementary education.* New York: Macmillan Publishing Company.

Gellert, E. (1962). Children's conceptions of the content and functions of the human body. *Genetic Psychology Monographs,* 65, 293-305.

Goldman, R. J., & Goldman, J. D. (1982). How children perceive the origin of babies and the role of mothers and fathers in procreation: A cross-national study. *Child Development,* 53, 491-504.

Good, R., Trowbridge, J., Demastes, S., Wandersee, J., Hafner, M., & Cummins, C. (Eds.). (1993). *Proceedings of the 1992 evolution education research conference.* Baton Rouge, LA: Louisiana State University.

Graeber, A., & Tirosh, D. (1988). Multiplication and division involving decimals: Preservice elementary teachers' performance and beliefs. *Journal of Mathematical Behavior,* 7, 263-280.

Greer, B. (1992). Multiplication and division as models of situations. In D. A. Grouws (Ed.), *Handbook of research on mathematics teaching and learning* (pp. 276-295). New York: Macmillan Publishing Company.

Grosslight, L., Unger, C., Jay, E., & Smith, C. (1991). Understanding models and their use in science: Conceptions of middle and high school students and experts. *Journal of Research in Science Teaching,* 28, 799-822.

## H

Hallam, R. N. (1970). Piaget and thinking in history. In M. Ballard (Ed.), *New movements in the study and teaching of history.* Bloomington: Indiana University Press.

Hallam, R. N. (1979). Attempting to improve logical thinking in school history. *Research in Education,* 21, 1-24.

Harrison, A., & Treagust, D. (1996). Secondary students' mental models of atoms and molecules: Implications for teaching science. *Science Education,* 80, 509-534.

Harrison, A., & Treagust, D. (2000). A typology of school science models. *International Journal of Science Education,* 22, 1011-1026.

Hellemans, A., & Bunch, B. (1991). *The timetables of science: A chronology of the most important people and events in the history of science.* New York: Simon & Schuster.

Heller, P. & Finley, F. (1992). Variable uses of alternative conceptions: A case study in current electricity. *Journal of Research in Science Teaching,* 29, 259-275.

Hembree, R., & Dessart, D. (1986). Effects of hand held calculators in precollege mathematics education: A meta-analysis. *Journal for Research in Mathematics Education,* 17, 83-89.

Henriques, L. (2002). Children's ideas about the weather: A review of the literature. *School Science and Mathematics,* 102, 202-215.

Herget, D. (Ed.). (1989). *The history and philosophy of science in science teaching.* Proceedings of the first international conference (Vols. 1-2). Tallahassee, FL: Florida State University.

Hess, R., & Torney, J. (1968). *The development of political attitudes in children.* Garden City, NY: Doubleday & Company.

Hiebert, J. (1992). Mathematical, cognitive, and instructional analyses of decimal fractions. In G. Leinhardt, R. Putnam, & R. Hattrup (Eds.), *Analysis of arithmetic for mathematics teaching* (pp. 283-322). Hillsdale, NJ: Lawrence Erlbaum Associates.

Hiebert, J., & Wearne, D. (1986). Procedures over concepts: The acquisition of decimal number knowledge. In J. Hiebert (Ed.), *Conceptual and procedural knowledge: The case of mathematics* (pp. 199-223). Hillsdale, NJ: Lawrence Erlbaum Associates.

Hills, S. (Ed.). (1992). *The history and philosophy of science in science education.* Proceedings of the second international conference (Vols. 1-2). Kingston, Ontario: Queen's University.

Huntley, M. A., Rasmussen, C. L., Villarubi, R. S., Sangtong, J., & Fey, J. T. (2000). Effects of standards-based mathematics education: A study of the Core-Plus Mathematics Project algebra and functions strand. *Journal for Research in Mathematics Education,* 31, 328-361.

## J

Jahoda, G. (1979). The construction of economic reality by some Glaswegian children. *European Journal of Social Psychology,* 19, 115-127.

Jahoda, G. (1981). The development of thinking about economic institutions: The bank. *Cahiers de Psychologie Cognitive,* 1(1), 55-73.

Johnson, C., & Wellman, H. (1982). Children's developing conceptions of the mind and brain. *Child Development,* 53(1), 222-234.

Johnson, P. (1998). Children's understanding of changes of state involving the gas state, Part 2: Evaporation and condensation below boiling point. *International Journal of Science Education,* 20, 695-709.

Jones, G., Carter, G., & Rua, M. (2000). Exploring the development of conceptual ecologies: Communities of concepts related to convection and heat. *Journal of Research in Science Teaching,* 37, 139-159.

Joyce, W., Little, T., & Wronski, S. (1991). Scope and sequence, goals, and objectives: Effects on social studies. In J. Shaver (Ed.), *Handbook of research on social studies teaching and learning* (pp. 321-331). New York: Macmillan Publishing Company.

Jungwirth, E. (1975). Preconceived adaptation and inverted evolution (a case study of distorted concept formation in high school biology). *Australian Science Teacher Journal,* 21, 95-100.

Jungwirth, E. (1987). Avoidance of logical fallacies: A neglected aspect of science education and science-teacher education. *Research in Science and Technological Education,* 5, 43-58.

Jungwirth, E., & Dreyfus, A. (1990). Identification and acceptance of a posteriori causal assertions invalidated by faulty enquiry methodology: An international study of curricular expectations and reality. In D. Herget (Ed.), *More history and philosophy of science in science teaching* (pp. 202-211). Tallahassee, FL: Florida State University.

Jungwirth, E., & Dreyfus, A. (1992). After this, therefore because of this: One way of jumping to conclusions. *Journal of Biological Education,* 26, 139-142.

## K

Kanari, Z., & Millar, R. (2004). Reasoning from data: How students collect and interpret data in science investigations. *Journal of Research in Science Teaching,* 41, 748-769.

Kaput, J. (1992). Technology and mathematics education. In D. A. Grouws (Ed.), *Handbook of research on mathematics teaching and learning* (pp. 515-556). New York: Macmillan Publishing Company.

Keil, F. (1989). *Concepts, kinds, and cognitive development.* Cambridge, MA: MIT Press.

Kesidou, S., & Duit, R. (1993). Students' conceptions of the second law of thermodynamics: An interpretive study. *Journal of Research on Science Teaching,* 30, 85-106.

Kesidou, S., Duit, R., & Glynn, S. (1995). Conceptual development in physics: Students' understanding of heat. In S. Glynn, & R. Duit (Eds.). *Learning science in the schools: Research reforming practice* (pp. 179-198). Mahwah, NJ: Lawrence Erlbaum Associates.

Kieran, C. (1992). The learning and teaching of school algebra. In D. Grouws (Ed.), *Handbook of research on mathematics teaching and learning* (pp. 390-419). New York: Macmillan Publishing Company.

Kieran, C., & Sfard, A. (1999). Seeing through symbols: The case of equivalent expressions. *Focus on Learning Problems in Mathematics,* 21(1), 1-17.

Kilpatrick, J., Swafford, J., & Findell, B. (Eds.) (2001). *Adding it up: Helping children learn mathematics.* Washington, DC: National Academy Press.

Klahr, D., Chen, Z., & Toth, E. (2001). Cognitive development and science education: Ships passing in the night or beacons of mutual illumination? In S. Carver, & D. Klahr (Eds.), *Cognition and Instruction: 25 Years of Progress.* Mahwah, NJ: Erlbaum.

Kouba, V., Brown, C., Carpenter, T., Lindquist, M., Silver, E., & Swafford, J. (1988). Results of the fourth NAEP assessment of mathematics: Numbers, operations, and word problems. *Arithmetic Teacher*, 35(8), 14-19.

Koulaidis, V., & Christidou, V. (1998). Models of students' thinking concerning the greenhouse effect and teaching implications. *Science Education,* 83, 559-576.

Krnel, D., Watson, R., & Glazar, S. (1998). Survey of research related to the development of the concept of "matter". *International Journal of Science Education,* 20, 257-289.

Kuhn, D. (1992). Thinking as argument. *Harvard Educational Review,* 62, 155-178.

Kuhn, D., Amsel, E., & O'Loughlin, M. (1988). *The development of scientific thinking skills.* San Diego, CA: Academic Press.

Kuhn, D., Black, J., Keselman, A., & Kaplan, D. (2000). The development of cognitive skills to support inquiry learning. *Cognition and Instruction,* 18, 495-523.

## L

Lazar, A., & Torney-Purta, J. (1991). The development of the subconcepts of death in young children: A short-term longitudinal study. *Child Development,* 62, 1321-1333.

Leach, J., Driver, R., Scott, P., & Wood-Robinson, C. (1992). *Progression in understanding of ecological concepts by pupils aged 5 to 16.* Leeds, UK: The University of Leeds, Centre for Studies in Science and Mathematics Education.

Leder, G. C., E. Pehkonen, & G. Toerner. (Eds.) (2002). *Beliefs: A hidden variable in mathematics education?* Dordrecht: Kluwer.

Lederman, N., & O'Malley, M. (1990). Students' perceptions of the tentativeness in science: Development, use, and sources of change. *Science Education,* 74, 225-239.

Lee, D. (n.d.) *Marks in the evolution of western thinking about nature.* Retrieved February 25, 2006, from http://www.sciencetimeline.net/index.htm.

Lee, O., Eichinger, D., Anderson, C., Berkheimer, G., & Blakeslee, T. (1993). Changing middle school students' conceptions of matter and molecules. *Journal of Research in Science Teaching,* 30, 249-270.

Lee, P. J. (1984). Historical imagination. In A. K. Dickinson, P. J. Lee, & P. J. Rogers (Eds.), *Learning history* (pp. 85-117). London: Heinemann Educational Books.

Lehrer, R., & Schauble, L. (2000). Modeling in mathematics and science. In R. Glaser (Ed.), *Advances in instructional psychology: Educational design and cognitive science* (Vol. 5) (pp. 101-159). Mahwah, NJ: Lawrence Erlbaum Associates.

Levstik, L. & Pappas, C. (1987). Exploring the development of historical understanding. *Journal of Research and Development in Education,* 21(1), 1-15.

Lewis, E., & Linn, M. (1994). Heat energy and temperature concepts of adolescents, adults, and experts: Implications for curricular improvements. *Journal of Research in Science Teaching,* 31, 657-677.

Licht, P. (1991). Teaching electrical energy, voltage and current: An alternative approach. *Physics Education,* 25, 272-277.

Linn, M., Clement, C., & Pulos, S. (1983). Is it formal if it's not physics? The influence of content on formal reasoning. *Journal of Research in Science Teaching,* 20, 755-776.

Linn, M., & Songer, N.B. (1991). Teaching thermo-dynamics to middle school students: What are appropriate cognitive demands? *Journal of Research in Science Teaching,* 28, 885-918.

Linn, M. & Swiney, J. (1981). Individual differences in formal thought: Role of cognitions and aptitudes. *Journal of Educational Psychology,* 73, 274-286.

## M

Mackey, J. (1991). Adolescents' social, cognitive, and moral development and secondary school social studies. In J. Shaver (Ed.), *Handbook of research on social studies teaching and learning.* New York: Macmillan Publishing Company.

Maloney, D. (1985). Charged Poles? *Physics Education,* 20, 310-316.

Markovits, Z., & Sowder, J. (1991). Students' understanding of the relationship between fractions and decimals. *Focus on Learning Problems in Mathematics,* 13(1), 3-11.

Marton, F. (1978). Phenomenography: Describing conceptions of the world around us. *Instructional Science,* 10, 177-200.

Matthews, M. (Ed.). (1991). *History, philosophy, and science teaching.* Selected readings. Toronto: OISE Press.

McDermott, L. & Shaffer, P. (1992). Research as a guide for curriculum development: An example from introductory electricity. Part I: Investigation of student understanding. *American Journal of Physics,* 60, 994-1003.

McLeod, D. (1992). Affect in mathematics education: A reconceptualization. In D. Grouws (Ed.), *Handbook of research on mathematics teaching and learning* (pp. 575-596). New York: Macmillan Publishing Company.

Mead, M., & Metraux, R. (1957). Image of the scientist among high-school students: A pilot study. *Science, 26*, 84-390.

Meadows, G., & Wiesenmayer, R. (1999). Identifying and addressing students' alternative conceptions of the causes of global warming: The need for cognitive conflict. *Journal of Science Education and Technology, 8*, 235-239.

Millar, R., & Lubben, F. (1996). Investigative work in science: The role of prior expectations and evidence in shaping conclusions. *Education, 3*, 28-34.

Miller, J. D., Scott, E. C., & Okamoto, S. (2006). Public acceptance of evolution. *Science, 313*, 765-766.

Mintzes, J., Trowbridge, J., Arnaudin, M., & Wandersee, J. (1991). Children's biology: Studies on conceptual development in the life sciences. In S. Glynn, R. Yeany, & B. Britton (Eds.), *The psychology of learning science* (pp. 179-202). Hillsdale, NJ: Lawrence Erlbaum Associates.

## N

National Research Council. (1996). *National Science Education Standards*. Washington, DC: National Academy Press.

Nelson, B., Aron, R., & Francek, M. (1992). Clarification of selected misconceptions in physical geography. *Journal of Geography, 91*, 76-80.

Newton, D., & Newton, L. (1992). Young children's perceptions of science and the scientist. *International Journal of Science Education, 14*, 331-348.

## O

Osborne, R., & Freyberg, P. (1985). Children's science. In R. Osborne & P. Freyberg (Eds.), *Learning in Science* (pp. 5-14). Auckland, NZ: Heinemann.

## P

Peck, D., & Jencks, S. (1981). Conceptual issues in the teaching and learning of fractions. *Journal for Research in Mathematics Education, 12*, 339-348.

Penner, D., Giles, N., Lehrer, R., & Schauble, L. (1997). Building functional models: Designing an elbow. *Journal of Research in Science Teaching, 34*, 125-143.

Philipp, R. A. (in press). Mathematics teachers' beliefs and affect. In F. Lester (Ed.), *Second handbook of research on mathematics teaching and learning*. Reston, VA: National Council of Teachers of Mathematics.

Philips, W. (1991). Earth science misconceptions. *Science Teacher, 58*, 21-23.

## R

Ramsey, P. (1986). Racial and cultural categories. In C. P. Edwards (Ed.), *Promoting social and moral development in young children* (pp. 78-101). Columbia University: Teachers College Press.

Ratcliffe, M. (1999). Evaluation of abilities in interpreting media reports of scientific research. *International Journal of Science Education, 21*, 1085-1099.

Resnick, L., Nesher, P., Leonard, F., Magone, M., Omanson, S., & Peled, I. (1989). Conceptual bases of arithmetic errors: The case of decimal fractions. *Journal for Research in Mathematics Education, 20*, 8-27.

Romberg, T., & Carpenter, T. (1986). Research on teaching and learning mathematics: Two disciplines of scientific inquiry. In M. Wittrock (Ed.), *Handbook of research on teaching* (pp. 850-873). New York: Macmillan Publishing Company.

Roseberry, A., Warren, B., & Conant, F. (1992). *Appropriating scientific discourse: Findings from language minority classrooms* (Working paper 1-92). Cambridge, MA: TERC.

Ross, J. A. (1988). Controlling variables: A meta-analysis of training studies. *Review of Educational Research, 58*, 405-437.

Rowell, J., & Dawson, C. (1984). Controlling variables: Testing a programme for teaching a general solution strategy. *Research in Science and Technological Education, 2*(1), 37-46.

Russell, T., Harlen, W., & Watt, D. (1989). Childrens' ideas about evaporation. *International Journal of Science Education, 11*, 566-576.

Russell, T., Longden, K., & McGuigan (1991). *Materials*. Primary Space Project Research Report. Liverpool, UK: Liverpool University Press.

Russell, T., & Watt, D. (1990). *Evaporation and condensation*. SPACE Project Research Report. Liverpool, UK: Liverpool University Press.

Rye, J., Rubba, R., & Wiesenmayer, R. (1997). An investigation of middle school students' alternative conceptions of global warming as formative evaluation of teacher-developed STS units. *International Journal of Science Education, 19*, 527-551.

## S

Sadler, P. (1987). Misconceptions in astronomy. In J. Novak (Ed.), *Proceedings of the second international seminar misconceptions and educational strategies in science and mathematics* (Vol. III, pp. 422-425). Ithaca, NY: Cornell University.

Sadler, P. (1998). Psychometric models of student conceptions in science: Reconciling qualitative studies and distractor-driven assessment items. *Journal of Research in Science Teaching, 35*, 265-296.

Salierno, C., Edelson, D., & Sherin, B. (2005). The development of student conceptions of the Earth-Sun relationship in an inquiry-based curriculum. *Journal of Geoscience Education, 53*, 422-431.

Sandoval, W. A., & Millwood, K. A. (2005). The quality of students' use of evidence in written scientific explanations. *Cognition and Instruction, 23*, 23-55.

Saxe, G., Becker, J., & Sadeghpour, M. (1989). Developmental differences in children's understanding of number word conventions. *Journal for Research in Mathematics Education, 20*, 468-488.

Schauble, L. (1996). The development of scientific reasoning in knowledge-rich contexts. *Developmental Psychology, 32*, 102-119.

Schauble, L., Klopfer, L. E., & Raghavan, K. (1991). Students' transition from an engineering model to a science model of experimentation. *Journal of Research in Science Teaching, 28*, 859-882.

Schoenfeld, A. (1985). *Mathematical problem solving*. New York: Academic Press.

Schoenfeld, A. (1989a). Explorations of students' mathematical beliefs and behavior. *Journal for Research in Mathematics Education, 20*, 338-355.

Schoenfeld, A. (1989b). Problem solving in context(s). In R. Charles & E. Silver (Eds.), *The teaching and assessing of mathematical problem solving* (pp. 82-92). Reston, VA: National Council of Teachers of Mathematics.

Schoenfeld, A. (1992). Learning to think mathematically: Problem solving, metacognition, and sense making in mathematics. In D. Grouws (Ed.), *Handbook of research on mathematics teaching and learning* (pp. 334-370). New York: Macmillan Publishing Company.

Schoenfeld, A. H. (2006). Mathematics teaching and learning. In P. A. Alexander & P. H. Winne (Eds.), *Handbook of educational psychology (2nd edition)*. Mahwah, NJ: Erlbaum.

Schug, M., & Birkey, C. (1985). The development of children's economic reasoning. *Theory and Research in Social Education*, XIII(1), 31-42.

Schug, M., & Lephardt, N. (1992, September). Development in children's thinking about international trade. *The Social Studies*, 207-211.

Schwarz, C., & White, B. (2005). Metamodeling knowledge: Developing students' understanding of scientific modeling. *Cognition and Instruction, 23*, 165-205.

Senk, S. (1989). Van Hiele levels and achievement in writing geometry proofs. *Journal for Research in Mathematics Education, 20*, 309-321.

Sere, M. (1985). The gaseous state. In R. Driver, E. Guesne, & A. Tiberghien (Eds.), *Children's ideas in science* (pp. 105-123). Milton Keynes, UK: Open University Press.

Shaffer, P., & McDermott, L. (1992). Research as a guide for curriculum development: An example from introductory electricity. Part II: Design of instructional strategies. *American Journal of Physics, 60*, 1003-1013.

Shelmit, D. (1984). Beauty and the philosopher: Empathy in history and classroom. In A. K. Dickinson, P. J. Lee, & P. J. Rogers (Eds.), *Learning history* (pp. 39-84). London: Heinemann Educational Books.

Shipstone, D. (1985). Electricity in simple circuits. In R. Driver, E. Guesne, & A. Tiberghien (Eds.), *Children's ideas in science* (pp. 33-51). Milton Keynes, UK: Open University Press.

Shipstone, D. M., Rhoeneck, C., Jung, W., Karrqvist, C., Dupin, J. S., Joshua, S., & Licht, P. (1988). A study of students' understanding of electricity in five European countries. *International Journal of Science Education, 10*, 303-316.

Shortland, M., & Warwick, A. (Eds.). (1989). *Teaching the history of science*. Oxford: Basil, Blackwell.

Silver, E. A., & Kenney, P. A. (2000). *Results from the seventh mathematics assessment of the National Assessment of Educational Progress*. Reston: VA: National Council of Teachers of Mathematics.

Smith, E., & Anderson, C. (1986, April). *Alternative conceptions of matter cycling in ecosystems.* Paper presented at the annual meeting of the National Association for Research in Science Teaching, San Francisco, CA.

Sodian, B., Zaitchik, D., & Carey, S. (1991). Young children's differentiation of hypothetical beliefs from evidence. *Child Development,* 62, 753-766.

Solomon, J. (1983). Learning about energy: How pupils think in two domains. *European Journal of Science Education,* 5, 49-59.

Solomon, J. (1985). Teaching the conservation of energy. *Physics Education,* 20, 165-170.

Solomon, J., Duveen, J., Scot, L., & McCarthy, S. (1992). Teaching about the nature of science through history: Action research in the classroom. *Journal of Research in Science Teaching,* 29, 409-421.

Sowder, J. (1988). Mental computation and number comparison: Their roles in the development of number sense and computational estimation. In J. Hiebert & M. Behr (Eds.), *Number concepts and operations in the middle grades* (pp. 182-197). Reston, VA: National Council of Teachers of Mathematics.

Sowder, J. (1992a). Making sense of numbers in school mathematics. In G. Leinhardt, R. Putnam, & R. Hattrup (Eds.), *Analysis of arithmetic for mathematics teaching* (pp.1-51). Hillsdale, NJ: Lawrence Erlbaum Associates.

Sowder, J. (1992b). Estimation and number sense. In D. Grouws (Ed.), *Handbook of research on mathematics teaching and learning* (pp. 371-389). New York: Macmillan Publishing Company.

Sowder, J. & Wheeler, M. (1989). The development of concepts and strategies used in computational estimation. *Journal for Research in Mathematics Education,* 20, 130-146.

Stavy, R. (1990). Children's conceptions of changes in the state of matter: From liquid (or solid) to gas. *Journal of Research in Science Teaching,* 27, 247-266.

Stone, L. C. (1986). International and multicultural education. In V. Atwood (Ed.), *Elementary social studies: Research as a guide to practice* (pp. 34-54). Washington, DC: National Council for the Social Studies.

Stylianidou, F., Ormerod, F., & Ogborn, J. (2002). Analysis of science textbooks' pictures about energy and pupils' readings of them. *International Journal of Science Education,* 24, 257-284.

Suydam, M. (1985). The shape of instruction in geometry: Some highlights from research. *Mathematics Teacher,* 78, 481-486.

## T

Threadgill-Sowder, J. (1984). Computational estimation procedures of young children. *Journal of Educational Research,* 77, 332-336.

Tiberghien, A. (1983a). Critical review on the research aimed at elucidating the sense that the notions of electric circuits have for students aged 8 to 20 years. In *Research in Physics Education: Proceedings of the first international workshop* (pp. 109-123). Paris: Centre National de la Recherche Scientifique.

Tiberghien, A. (1983b). Critical review of the research aimed at elucidating the sense that notions of temperature and heat have for students aged 10 to 16 years. In *Proceedings of the first international workshop research on physics education* (pp. 73-90). Paris: Editions du CNRS.

Tiberghien, A. (1985). Heat and temperature: The development of ideas with teaching. In R. Driver, E. Guesne, & A. Tiberghien (Eds.), *Children's ideas in science* (pp. 66-84). Milton Keynes, UK: Open University Press.

Tierney, C., & Nemirovsky, R. (1991, Fall). Children's spontaneous representations of changing situations. *Hands on!,* 7-10.

Thompson, A. G. (1992). Teachers' beliefs and conceptions: A synthesis of the research. In D.A. Grouws (Ed.), *Handbook of research on mathematics teaching and learning* (pp. 127-146). New York: Macmillan.

Tomasini, G., & Balandi, P. (1987). Teaching strategies and children's science: An experiment on teaching "hot and cold." In J. Novak (Ed.), *Proceedings of the second international seminar "misconceptions and educational strategies in science and mathematics"* (Vol. II, pp. 158-171). Ithaca, NY: Cornell University.

Treagust, D., Chittleborough, G., & Mamiala, T. (2002). Students' understanding of the role of scientific models in learning science. *International Journal of Science Education,* 24, 357-368.

Trowbridge, J., & Mintzes, J. (1985). Students' alternative conceptions of animals and animal classification. *School Science and Mathematics,* 85, 304-316.

Turiel, E. (1983). *The development of social knowledge.* Cambridge: Cambridge University Press.

## V

Van Hiele, P. (1986). *Structure and insight.* Orlando, FL: Academic Press.

Viennot, L. (2001). Reasoning in physics: The part of common sense. Dordrecht: Kluwer Academic Publishers.

Vosniadou, S. (1991). Designing curricula for conceptual restructuring; lessons from the study of knowledge acquisition in astronomy. *Journal of Curriculum Studies,* 23, 219-237.

## W

Waterman, M. (1983). Alternative conceptions of the tentative nature of scientific knowledge. In J. Novak (Ed.), *Proceedings of the international seminar on misconceptions in science and mathematics* (pp. 282-291). Ithaca, NY: Cornell University.

Watts, M. (1983a). Some alternative views of energy. *Physics Education,* 18, 213-217.

Wearne, D., & Hiebert, J. (1988). Constructing and using meaning for mathematical symbols: The case of decimal fractions. In J. Hiebert & M. Behr (Eds.), *Number concepts and operations in the middle grades* (pp. 220-235) Reston, VA: National Council of Teachers of Mathematics.

Wearne, D., & Hiebert, J. (1989). Cognitive changes during conceptually based instruction on decimal fractions. *Journal of Educational Psychology,* 81, 507-513, ERIC No. ED 254 409.

Wellman, H. M., & Johnson, C. (1982). Children's understanding of food and its functions: A preliminary study of the development of concepts of nutrition. *Journal of Applied Developmental Psychology,* 3, 135-148.

Wirszup, I. (1976). Breakthroughs in the psychology of learning and teaching geometry. In J. Martin & D. Bradbard (Eds.), *Space and geometry: Papers from a research workshop* (pp. 75-97). Athens, GA: University of Georgia, Georgia Center for the Study of Learning and Teaching Mathematics. (ERIC Reproduction Service No. ED 132 033).

Wiser, M. (1986). *The differentiation of heat and temperature: An evaluation of the effect of microcomputer teaching on students' misconceptions.* Technical report. Cambridge, MA: Harvard Graduate School of Education.

Wiser, M. (1988). The differentiation of heat and temperature: History of science and novice-expert shift. In S. Strauss (Ed.), *Ontogeny, phylogeny, and historical development.* Norwood, NJ: Ablex Publishing Corporation.

Wollman, W. (1977a). Controlling variables: Assessing levels of understanding. *Science Education,* 61, 371-383.

Wollman, W. (1977b). Controlling variables: A neo-Piagetian developmental sequence. *Science Education,* 61, 385-391.

Wollman, W., & Lawson, A. (1977). Teaching the procedure of controlled experimentation: A Piagetian approach. *Science Education,* 61, 57-70.

## Z

Zan, R., Brown, L., Evans, J., & Hannula, M. S. (2006). Affect in mathematics education: An introduction. *Educational Studies in Mathematics,* 63, #2, 113-121

Zeidler, D. L. (1997). The central role of fallacious thinking in science education. *Science Education,* 81, 483-496.

Zimmerman, C. (2000). The development of scientific reasoning skills. *Developmental Review,* 20, 99-149.

Zimmerman, C. (2005, August). *The development of scientific reasoning skills: What psychologists contribute to an understanding of elementary science learning.* Paper prepared for the National Research Council, Committee on Science Learning Kindergarten through Eighth Grade. Illinois State University.

# ACKNOWLEDGMENTS

## PROJECT 2061 STAFF

**Michael Anderson,** Senior Business Analyst

**Jonah Ben-Joseph,** Writer

**Abigail Burrows,** Senior Project Coordinator

**George DeBoer,** Deputy Director of Project 2061

**Arhonda Gogos,** Research Associate

**Barbara Goldstein,** Program Associate

**Serita Henderson,** Program Operations Coordinator

**Cari Herrmann Abell,** Research Associate

**Cheryl Jackson,** Financial Analyst

**Sofia Kesidou,** Consultant to Project 2061*

**Mary Koppal,** Communications Director

**Kristen Lennon,** Research Associate

**Lester Matlock,** Administration Director

**Francis Molina,** Technology Director

**Kathleen Morris,** Senior Program Associate

**David Peery,** Technology Specialist/Network-MS Administrator

**Thomas Regan,** Research Associate

**Jo Ellen Roseman,** Director of Project 2061

**Brian Sweeney,** Technology Applications Developer

**Catherine Tramontana,** Senior Editor

**Stedman (Ted) Willard,** Senior Program Associate**

* Sofia Kesidou directed work on *Atlas 2* from Fall 2001 through Spring 2005 in her role as Associate Program Director and has continued to provide important contributions from her new home in Sydney, Australia.

** Although many staff and colleagues contributed extensively to the work reflected in *Atlas 2*, these maps would certainly not exist were it not for the monumental efforts of Ted Willard.

## REVIEWERS*

**Jeannette E. Allen,** Senior Science Education Specialist, Science Systems and Applications, Inc., at NASA Goddard Space Flight Center

**Ben Almassi,** Graduate Student, Philosophy, University of Washington

**Charles Anderson,** Professor of Teacher Education, Michigan State University

**Monica Aufrecht,** Graduate Student, Philosophy, University of Washington

**Ann Benbow,** Director, Education and Outreach, American Geological Institute

**Rebecca Boger,** Deputy Director, The GLOBE Program

**Mary Ellen Bowden,** Senior Research Historian, Chemical Heritage Foundation

**Leah Bricker,** Graduate Student, Cognitive Studies in Education, School of Education, University of Washington

**Audrey Champagne,** Professor Emerita, Educational Theory and Practice, University at Albany, State University of New York

**Vincent Childress,** Associate Professor of Technology Education, North Carolina A & T University

**C. Herbert Clemens,** Professor of Mathematics, Ohio State University

**Brian Coppola,** Professor of Chemistry, University of Michigan

**Martin Creel,** Acting Director, Division of Accelerated and Enriched Instruction, Montgomery County Public Schools, MD

**Dolores D'Angelo,** Program Coordinator, Communications Arts Program, Montgomery Blair High School, Montgomery County Public Schools, MD

**Tim Eichler,** Physical Scientist, Office of Oceanic & Atmospheric Research, National Oceanic & Atmospheric Administration

**Arthur Eisenkraft,** Professor, Graduate College of Education, University of Massachusetts Boston

**Bradford Findell,** Assistant Professor, Mathematics and Science Education, University of Georgia

**David Fortus,** Assistant Professor of Teacher Education, Michigan State University and The Weizmann Institute for Science

**Owen Gingerich,** Professor Emeritus, History of Science, Harvard University and Senior Astronomer Emeritus, Smithsonian Astrophysical Observatory

**Ron Gird,** Outreach Program Manager, National Weather Service, National Oceanic & Atmospheric Administration

**Fred M. Goldberg,** Director, Center for Research in Mathematics and Science Education, San Diego State University

**Kenneth J. Heller,** Professor, Associate Head (Physics), University of Minnesota

**Patricia M. Heller,** Associate Professor Emeritus, Department of Curriculum and Instruction, University of Minnesota

**Pamela Hines,** Senior Editor, *Science*, AAAS

**Julie Homchick,** Graduate and Teaching Assistant in Communications, University of Washington

**Katrina Kelner,** Deputy Editor, Life Sciences, *Science*, AAAS

**Keith Kester,** Professor of Chemistry, Colorado College

**Sharon Kingsland,** Department Chair, History of Science and Technology, Johns Hopkins University

**Joseph Krajcik,** Professor, Science Education, Associate Dean for Research, Co-Director, Center for Curriculum Materials in Science, University of Michigan

**Deanna Kuhn,** Professor of Psychology and Education, Teachers College, Columbia University

**Anton Lawson,** Professor of Biology Education and Cognitive Sciences, Arizona State University

**John Layman,** Profesor Emeritus, Curriculum and Instruction, University of Maryland

**Rhonda Levine,** Professor of Sociology, Colgate University

**Julie Libarkin,** Assistant Professor, Geological Sciences, Michigan State University

**Mark McCaffrey,** CIRES Education and Outreach, University of Colorado, Boulder

**William McComas,** Parks Family Professor of Science Education, University of Arkansas

**Donald McKinney,** Education Consultant, Chemical Heritage Foundation

**Erin McLeary,** Curator of Exhibitions, Chemical Heritage Foundation

**Douglas McLeod,** Professor Emeritus of Mathematics, San Diego State University

**Mark Michalovic,** Education Consultant, Chemistry, Chemical Heritage Foundation

**Forrest Mims,** Instrument Designer, Science Writer and Independent Science Consultant, Geronimo Creek Observatory

**George Nelson,** Director, Science, Mathematics and Technology Education, Western Washington University

**Pamela Newberry,** Director of Curriculum for Pre-Engineering, Project Lead The Way

**Frank Niepold,** Climate Education Coordinator, UCAR Visiting Scientist, Climate Program Office, National Oceanic & Atmospheric Administration

**Gilbert Omenn,** Executive Vice President for Medical Affairs, University of Michigan Health System and Chair, AAAS Board of Directors, 2006-2007

**Gregory Pearson,** Program Officer, National Academy of Engineering

**Lawrence Principe,** Professor of History of Alchemy/Chemistry, Johns Hopkins University

**Edward F. Redish,** Professor of Physics, University of Maryland

**Judy A. Reeves,** Albert Einstein Distinguished Educator Fellow, Office of Education, National Oceanic & Atmospheric Administration

**Cheryl Rose,** Senior Program Director, Mathematics, Maine Mathematics and Science Alliance

**Louis Rosenblatt,** Science Teacher, Park School of Baltimore

**Matthew Rossi,** Research Analyst, IQWST, Northwestern University

**F. James Rutherford,** Founding Director, Project 2061, AAAS

**Linda Schmidt,** Associate Professor, Mechanical Engineering, University of Maryland

**Joseph Shewmaker,** Albert Einstein Distinguished Educator Fellow, Office of Education, National Oceanic & Atmospheric Administration

**Duncan Sibley,** Director, Center for Integrative Studies-General Science, Michigan State University

**Carol Smith,** Associate Professor, Department of Psychology, University of Massachusetts Boston

**Peg Steffen,** Education Coordinator, National Ocean Services, National Oceanic & Atmospheric Administration

**Luli Stern,** Consultant to Project 2061

**Andrew M. Sugden,** International Managing Editor, *Science*, AAAS

**Phil Szuromi,** Supervisory Senior Editor, *Science*, AAAS

**John Teibault,** Education Manager, Roy Eddleman Institute for Interpretation and Education, Chemical Heritage Foundation

**Albert Van Helden,** Professor Emeritus, History, Rice University

**Donald Waldman,** Richard M. Kessler Professor of Economic Studies, Colgate University

**Diana Wearne,** Professor Emeritus, School of Education, University of Delaware

**Peyton West,** Senior Program Associate, Dialogue on Science, Ethics, and Religion, AAAS

**Soren Wheeler,** Consultant to Project 2061

**Linda D. Wilson,** Consultant to Project 2061

**Paula N. Wilson,** Consultant to Project 2061

**Jake Yeston,** Associate Editor, *Science*, AAAS

**John Zola,** Social Studies Teacher, New Vista High School, Boulder, CO

* A special thanks to the many anonymous reviewers who provided feedback on maps through Frances Lawrenz, Associate Vice President for Research, University of Minnesota, as part of an external evaluation conducted by The Evaluation Center at Western Michigan University.

## EDITORIAL

**Peggy Dillon**

**Toni Maltagliati**

**Amani Rushing,** Directorate of Education and Human Resources, AAAS

## DESIGN AND PRODUCTION

**Greg Berger,** Kessler Design Group

**Suzanne Kesler,** Kessler Design Group

**Ethel Kessler,** Kessler Design Group

# CREDITS

Page viii: Drawing by Andrew Ahlgren.

Page xii: © Anglo-Australian Observatory/David Malin Images.

Page 2: © Estate of Alexander Calder/Artists Rights Society (ARS), New York. Photograph Copyright: © 1998: Whitney Museum of American Art, New York.

Page 6: © 2007 ScienceCartoonsPlus.com.

Page 10: From *Science,* vol. 170 (October 23, 1970).

Page 12: © 2007 ScienceCartoonsPlus.com.

Page 14: From *Science,* vol. 197 (August 26, 1977).

Page 18: Gift of the artist. © Board of Trustees, National Gallery of Art, Washington.

Page 20: © The *New Yorker* Collection 2006, Lee Lorenz from cartoonbank.com. All Rights Reserved.

Page 28: John Hay Whitney Collection, © Board of Trustees, National Gallery of Art, Washington.

Page 32: The Far Side® by Gary Larson © 1981 FarWorks, Inc. All Rights Reserved. The Far Side® and the Larson® signature are registered trademarks of FarWorks, Inc. Used with permission.

Page 34: Princeton University Art Museum. Arthur M. Sackler Collection, acc. no.67-2a. Photo Credit: Bruce M. White © Photos: Trustees of Princeton University.

Page 42: Paul Mellon Collection, © Board of Trustees, National Gallery of Art, Washington.

Page 44: © The *New Yorker* Collection 1986, J. B. Handelsman from cartoonbank.com. All Rights Reserved.

Page 52: Courtesy of the John Deere Art Collection, Deere & Company, Moline, IL.

Page 62: Hirshhorn Museum and Sculpture Garden, Smithsonian Institution, Gift of the Jospeh H. Hirshhorn Foundation, 1966. Lee Stalsworth, Photographer.

Page 68: © 2007 ScienceCartoonsPlus.com.

Page 70: Photographs courtesy of the University of Oklahoma Library and the Harvard College Observatory.

Page 72: © 2007 ScienceCartoonsPlus.com.

Page 78: © 2007 ScienceCartoonsPlus.com.

Page 90: Girard Collection, Museum of International Folk Art (DCA), Santa Fe, New Mexico. Photo by Michel Monteaux.

Page 100: Page from a Leonardo da Vinci notebook, 15-16th centuries.

Page 106: © 2007 ScienceCartoonsPlus.com.

Page 112: © 2007 ScienceCartoonsPlus.com.

Page 114: The Metropolitan Museum of Art, The Berggruen Klee Collection, 1984 (1984.315.33). Photograph © 1986 The Metropolitan Museum of Art.